Planning and Managing Major Construction Projects

Planning and Managing Major Construction Projects

A Guide for Hospitals

Deborah J. Rohde
Lawrence D. Prybil
William O. Hochkammer

Health Administration Press Perspectives
Ann Arbor, Michigan
1985

Library of Congress Cataloging in Publication Data

Rohde, Deborah J.
Planning and managing major construction projects.

Bibliography: p.
Includes index.
1. Hospitals—Design and construction. 2. Hospitals—Planning. I. Prybil, Lawrence D. II. Hochkammer, William O. III. Title. [DNLM: 1. Hospital Design and Construction. 2. Hospital Planning. WX 140 R737p]
RA967.R59 1985 362.1′1′0682 84-28997
ISBN 0-910701-01-6 (pbk.)

Health Administration Press
School of Public Health
The University of Michigan
1021 East Huron Street
Ann Arbor, Michigan 48109
(313) 764-1380

Health Administration Press Perspectives is an imprint of Health Administration Press dedicated to books and other material of timely and special interest for health care practitioners.

Contents

Preface

During 1982 and 1983, Sisters of Mercy Health Corporation developed *A Guide for Planning and Managing Major Construction Projects* to provide assistance to its senior hospital management, the hospital governance, and others in connection with major construction projects. The *Guide* was based on the substantial experience of Sisters of Mercy Health Corporation with these projects since its formation in 1976. The *Guide* has been utilized by the operating divisions of Sisters of Mercy Health Corporation and has proven to be quite useful.

Sisters of Mercy Health Corporation currently owns and operates 23 health care facilities in three states—Michigan, Iowa, and Indiana—with a combined total licensed bed complement of 5,780 beds. In addition, the Corporation operates six hospitals under management contracts in Iowa and New York. Sisters of Mercy Health Corporation is sponsored by the Detroit Province of the Sisters of Mercy, a Roman Catholic order which has been active in the health care field for more than 150 years. Many of the Sisters of Mercy Health Corporation facilities have been or are involved in major construction projects.

Sisters of Mercy Health Corporation has had sufficient experience with the *Guide* to believe that it may be of value to the health care industry and has thus decided to make it available for publication. This version of the *Guide* has been adapted for general use. We hope that it will serve the same purposes for the industry as it has served for Sisters of Mercy Health Corporation, namely to: (1) Provide an understanding to hospital chief executive officers and other administrative staff regarding the complexity of major construction projects; (2) Outline the basic phases of planning and implementing major hospital construction projects; (3) Indicate the probable governance approval points in the capital planning and implementation process; (4) Explain the roles of hospital staff and consultants on the project team; (5) Define technical terms associated with a construction program; and (6) Suggest strategies, methods, and approaches for project organization, team structure, task implementation, scheduling, and contracting.

In addition to senior hospital management, the *Guide* may be appropriate for use by the governing board, medical staff leaders, department directors, and legal counsel.

This *Guide* is intended to provide an overview on the subject. In-depth information is not included, particularly in connection with issues which are important to, but collateral to, the main subject. More in-depth information can be obtained from materials listed in the supplemental reading list.

The conception of the original *Guide* and its initial outline were developed by members and staff of a Sisters of Mercy Health Corporation Task Force on Construction Planning and Management which consisted of Mr. Larry S. Anderson, Ms. Judith K. Call, Mr. Charlie Carpenter, Mr. Charles Dreher, Sr. Frances M. Gerhard, R.S.M., Mr. Elliott Guttman, Mr. William O. Hochkammer, Esq., Mr. Lynn Nellenbach, Mr. Lawrence D. Prybil, and Mr. P. Whitney Spaulding.

This group has also contributed significant time and effort reviewing and revising the original *Guide* during its drafting stage. The firm of Health Facilities Corporation, and particularly Ms. Deborah J. Rohde, Senior Vice President, had principal responsibility for drafting the original *Guide* based upon the concepts and outline developed by the Task Force.

EDWARD J. CONNORS
President

April, 1984

I

Introduction

Planning and managing a major construction project is a multifaceted interdisciplinary task which requires a significant time commitment, attention to detail, and the integration of many complex elements. Due to the complexity of the construction planning and implementation process, any document which attempts to outline and describe the process must necessarily be somewhat detailed and lengthy.

The length of this *Guide*, however, must not be construed to represent that this is a definitive and complete text. The amount of information in this *Guide* was limited to a reasonably manageable level for review and reference purposes. Other source material should be consulted for additional detail or specific project issues.

The organization of the book is based upon the chronology of an actual project, and proceeds from a discussion of the general to the specific. First, certain general planning activities which should precede the specific project planning efforts are addressed. Chapter III, Overview of Project Phases, provides a general summary of the entire construction planning and implementation process with a brief description of each phase in the process. This chapter may serve as the best reference for overview purposes. Then the roles of the project team members and the managing of the team are discussed. Finally, chapters V through XI explain the specific project phases in detail. Following the body of the document are a supplemental reading list and a glossary of key terms.

II

Before the Project Is Initiated

The initiation of project activities is really the beginning of the end. It is the implementation of plans and strategies which have been developed on an ongoing basis by the institution. Early planning activities are crucial to the orderly development of the project planning and management process, since the early planning work will serve as the foundation upon which the construction project is developed. The following sections discuss important considerations which should be addressed before the project is initiated, including strategic planning, program development, long-range financial planning, and physical facilities planning.

The Strategic Plan

Planning is a learning process which can enhance the decision-making process of an institution through the development of the decision makers' awareness, sensitivity, and skills. Strategic planning provides the foundation and framework for all other governance and management activities. The activities of the strategic planning function are as follows:*

1. Review, revise philosophy/mission
   - Review philosophy/mission
   - Update philosophy/mission
   - Document, publish philosophy/mission

IGMP, Integrated Governance and Management Process^CTM Conceptual Design, Approved by the Sisters of Mercy Health Corporation Board of Trustees, April 11, 1980.

2. Develop role, goals, objectives, alternate strategies
   — Assess external environment
   — Assess internal environment
   — Define current role
   — Identify long-term goals
   — Identify intermediate term objectives
   — Identify alternate strategies
   — Document and promulgate role, goals, objectives, strategies

It is essential that a strategic plan be developed prior to initiating planning of a specific construction project. That is, the strategic plan serves as the foundation for the specific construction project. The project should be the implementation of one or more of the objectives of the strategic plan and should be a rational outcome of the strategic planning process. The plan should have justified a particular project alternative in terms of the following plan development information:

— Service area population
— Demographic characteristics
— Health status characteristics
— Health resource inventory and utilization (service area)
— Analysis of the hospital (utilization, medical staff, education)
— Financial resource characteristics

Some of the information will require updating and refinement once specific project planning is initiated.

The development of a strategic plan before starting a specific capital project will help insure that:

1. The project is based on broad institutional goals
2. The project is planned for verified service volumes
3. The project is designed to fit reasonably within the hospital's debt capacity
4. Facility changes are planned consistent with the hospital's program development

PROGRAM DEVELOPMENT

Program development entails three plans to achieve long-term goals: a strategic plan, an operational plan, and an organizational plan.

The activities of the program development function are as follows:*

1. Develop strategic plan
   — Assess alternatives
   — Estimate resource requirements
   — Select optimal strategy
   — Define program/service structure
   — Compile document/strategic plan
   — Document methodology and assumptions
2. Develop operational plan
   — Select fiscal year targets
   — Determine priorities
   — Define program/service structure and project utilization
   — Establish basis for evaluation
   — Estimate resource requirements
   — Compile operational plan
   — Document/publish plan
3. Develop organizational plan
   — Assess organizational characteristics
   — Assess ability to implement plan
   — Develop organizational plan
   — Define accountability relationships
   — Delegate/assign authority
   — Compile organizational plan
   — Promulgate organizational plan

The planning of project activities should come after program development so that the goals of the project are consistent with the aims of the hospital's program development. During the early planning activities of the project, the program work already accomplished may be refined and elaborated.

Early program development enhances the project planning phase. However, it should clarify, among others, the following points:

1. Comprehensiveness of clinical/nonclinical program being discontinued, initiated, or expanded (this has attendant space and technology requirements)
2. Effect on staffing of discontinued, initiated, or expanded programs (this could affect financial feasibility of project)
3. Competitive programs in health service area

---

*Ibid.

4. Service volumes related to programmatic changes (this affects space requirements, financial feasibility, and 1122/CON planning)

LONG-RANGE FINANCIAL PLANNING AND DEBT CAPACITY

Prior to embarking on project planning activities, an institution should understand the financial parameters within which the project planning must be confined in order to make the most constructive use of work in developing CON applications and architectural plans. Long-range financial planning should include an assessment of:

1. Financial summary and key ratios
2. Patient revenue by payer and type
3. Analysis of profit and loss overall and by product line
4. Review of existing financial structure
5. Historical and projected cost per unit of service
6. Planned sources and uses of funds for program development

In connection with its strategic planning and program development activities, the hospital should be conducting long-range financial planning, the results of which may serve in turn to modify the objectives of the strategic and program plans. This may be a push/pull process depending on the institution's financial character.

A long-range financial plan should include:

1. A long-range projection of future operating and capital requirements for a five year planning horizon
2. Analysis of historical and current trends in utilization and prices
3. Projections of expected changes in volume, staffing, capital, and other needs related to changes in programs and proposed projects
4. Analysis and accommodation of such identifiable constraints as reasonable rate increases, industry averages of the capital marketplace, restrictions associated with existing debt covenants, and applicable financial guidelines

Based upon projected financial statements, projected cash flows, and the quantified result of the future operating and capital requirements, the hospital may determine whether its proposed construction plans are within or exceed the institution's debt capacity. Should the plans be in excess of the hospital's debt capacity, priorities must be set, and the projections must be revised until the capital plans fit within the established constraints.

As a result of the federal prospective payment system legislation in 1983, the hospital industry is undergoing a transition from cost-based reimbursement to fixed-price reimbursement per diagnostic related group (DRG) for Medicare patients. Individual state's Medicaid programs as well as certain Blue Cross payment mechanisms may also be under a DRG type reimbursement system. Until 1986, capital costs will continue to be reimbursed as a direct passthrough. Subsequently, capital reimbursement will be according to a fixed-price formula, most likely as a component of total operating expenses.

A hospital's long-range financial plan must address the institutions' position under full implementation of the prospective payment systems. The current and projected profit or loss status of the hospital's product line should be identified on a specific product basis, i.e., by DRG. Such an evaluation could suggest classic make/buy decisions for specific products. This will have implications for the hospital's strategic plan and capital construction requirements. Long-range financial planning may indicate development potential for divestitures, operating joint ventures, and diversification. The long-range financial plan and strategic plan are therefore integrally related.

MASTER PHYSICAL FACILITIES PLANNING

After a hospital has prepared a strategic plan and has conducted program development activities, the hospital may find that it must increase its physical capacity, correct major code violations, and/or improve the functional configuration of the facility to implement the program development plans. A debt capacity analysis should have been conducted to verify that the hospital can afford a project of a given size. The next step would be to conduct master physical facilities planning.

Master physical facilities planning includes:

— A master site plan

— A master facility plan

— A code analysis

— An engineering systems analysis

One or more of the above physical planning efforts may be an ongoing activity of the hospital's strategic planning or may have already been accomplished in some other manner. If so, at the point that the specific construction project is conceived, the master physical facility planning issues may require only confirmation.

Master Site Plan

The master site plan should represent, at a conceptual level, the most current thinking regarding the ultimate development of the hospital's site. This plan includes the current land use (buildings, roads, access) as well as projected future use. The future plans note proposed building expansions and demolitions, proposed parking, potential land acquisition, and such proposed nonacute care development as medical office buildings or extended care facilities. The master site plan should also indicate such physical constraints as flood plains, retention ponds, and availability of utilities and associated connections. The master site plan should:

1. Provide a road map for the future development of the site to assure that the hospital will not be precluded from a particular development plan by limitations of the campus
2. Identify the site, zoning, building, and land use constraints and other restrictions of a legal nature which could limit or otherwise affect proposed program development, so that either site or program alternatives can be evaluated

Master Facility Plan

Like the master site plan, the master facility plan should represent, on a conceptual basis, the most current thinking regarding the ultimate development of the facility, but relative to the location of key departments and functions within the current and future buildings. The master facility plan should evaluate and identify:

1. Current and appropriate future functional zones for patient and nonpatient care
2. Current and appropriate future functional relationships between key departments
3. Growth plans for patient, ancillary, and support departments
4. Current and future horizontal and vertical transportation systems within buildings and appropriate traffic patterns for flow of supplies, patients, staff, and public

The function of the master facility plan is to provide a long-range direction to the internal physical development of the facility so that each construction phase is part of an orderly, well thought out development plan.

Code Analysis

On a periodic basis and prior to initiating any major construction project, a hospital should conduct an analysis of the condition of its physical facilities relative to life safety, fire, and other building codes. This is important for a number of reasons:

- — Codes are constantly revised
- — Correcting code violations serves as project justification in the 1122/CON approval process
- — Including code compliances in the project design is generally required to secure a building permit and approval of the fire marshall
- — Code compliances can have a significant effect on the final cost of the project
- — Code compliances may be required to secure an occupancy permit.

The degree to which code compliances will be enforced by the various agencies will become apparent later during the agency design reviews. However, having an early knowledge of code deficiencies is helpful, so that budget contingencies may be developed to remedy the deficiencies, as required, and the effect of such remedial action on program and function may thus be accommodated early in the project planning process.

Depending on the reviewing agency, correction of all code violations may not be enforced or code compliances may thus be negotiable. Should code waivers be received, adequate documentation of the waivers should be secured.

Engineering Systems Analysis

Nearly every major construction project, excluding complete replacement facilities, will entail certain upgrading, expansion, or renovation of the existing mechanical, electrical, plumbing, heating, ventilating, and air-conditioning systems. Prior to planning a specific construction project, a hospital should have a basic understanding of the capacity and condition of its plant engineering systems, for the following reasons:

1. Upgrading engineering systems may be required to comply with codes
2. Engineering systems average 40–55 percent of the cost of a project and therefore a major systems upgrade will significantly impact the project budget

3. Increased boiler/electrical/etc. capacity may be necessary to support a proposed expansion of programs or services
4. Upgrading or expanding a particular engineering system may have associated space requirements which must be accommodated in the proposed project

Generally, the detailed engineering design of a construction project does not occur until 1122/CON approval is received and schematic design is complete. But early attention to the basic engineering requirements of the plant could save time and money during the design of the project.

III

Overview of Project Phases

This chapter starts with a general discussion of the various phases in planning and developing a capital construction project. Later chapters address the specific phases in detail. The recommended governing board approval points are integrated in the detailed discussions of the project phases, as well as summarized below.

Governing Board Approvals

An institution's strategic plan should be approved by its governing board. There are either three or four recommended governing board approval points for a specific construction project, depending on whether long-term debt is involved or not. The governing body approval authority may rest with the full governing board or be delegated to an executive, finance, or project committee of the board, depending on the nature of the particular approval and on the hospital's governing documents and practices:

1. Approval of the project in concept prior to filing CON application
2. Approval to award the architectural/engineering contract prior to initiating schematic design
3. Approval of long-term financing prior to signing a financing commitment
4. Approval to award construction contracts

The intervening time period between governing board approval points could be lengthy and it is therefore prudent to keep the governing board informed on the status of the project through progress reports to a committee of the board. The hospital administration may wish to make presentations at milestone dates in the design process. For example, upon completion of the schematic design and the schematic cost estimate, a progress report may be made to verify that the project is on budget and on schedule, and a refined financing plan, as proposed, may be reviewed.

Program Stage and CON Approval

The detailed program development for the project and CON approval stage are the first steps in the project planning activities subsequent to the completion of the following work tasks and analyses previously described:

— Strategic plan
— Program development
— Long-range financial plan and debt capacity
— Master physical facilities planning

The detailed program development task and the processing of CON approval are discussed together since these should be concurrent planning activities. During the detailed program development, the specific project definition should emerge based upon space and functional needs, code requirements, future demand for services, schedule, and debt capacity. These planning criteria are similar to those required in developing a CON application and the CON approval process has attendant guidelines which should be referenced during the program development process.

The program development activity should result in:

1. Projected utilization for a five to seven year planning horizon for inpatient services, outpatient services, and major ancillary departments
2. Preliminary program narrative which includes discussion of function, operating systems, departmental relationships, and internal circulation patterns
3. A projection of future space requirements by department, based upon projected utilization and program narrative, to include estimates for new construction and renovated space
4. Block drawings based upon the square footage projections which indicate spaces and relationships for each department, but not necessarily spaces within departments
5. Master project schedule for planning, design, financing, and construction
6. Construction cost estimates based upon the preliminary program narrative, block drawings, and schedule

7. Project budget for equipment, fees, contingencies, and administrative expenses based upon the estimated construction cost
8. Plan of financing based upon the project budget and schedule
9. Projected statements of revenue and expenses and cash flow to verify that the project fits within the hospital's debt capacity

As each of the above tasks are completed, the results should be analyzed in the context of government planning criteria and guidelines which may include:

1. Approved bed need methodology
2. Service area needs and accessibility and availability of services
3. Square footage per bed or procedure
4. Cost per square foot
5. Cost per equivalent patient day
6. Rate increases

As stated previously, the program development stage may involve a push/pull process as needs are compared with financial and government planning constraints. The details of the program development work tasks and the CON approval process are discussed in later chapters.

Design

The various design phases in developing a capital construction project are as follows:

1. Block drawings based on program narrative
2. Schematic design
3. Design development
4. Construction documents or working drawings
5. Record drawings or as-built drawings

Block drawings indicate blocks of spaces for departments. They are primarily developed to identify and test the interdepartmental relationships and functional configurations between major clinical and support areas. Internal horizontal and vertical circulation patterns and systems as well as space zoning concepts (i.e., patient, staff, and public zones) are addressed on a general level in block drawings. The new and/or renovated building in relation to its site is analyzed and such issues as utilities, site access, and vehicular campus traffic patterns are studied. This stage of design should be a confirmation of the proposed facility development as indicated in the hospital's master facility plan and master plan site. A preliminary program narrative and block drawings are generally required for the CON approval process. Upon receipt of the CON approval, the program narrative is often refined and expanded prior to initiating schematic design.

Schematic design is generally not initiated until CON approval is received since this design phase could involve a significant fee commitment. Schematic design is the first architectural layout of spaces within and between departments, including corridors, offices, closets, and other storage areas. Schematic design is generally drawn at 1/16″ or 1/32″ scales, i.e., a smaller scale than subsequent phases of design. It would include the first structural analysis and engineering calculations for the building.

Design development is the first stage when department managers provide detailed input regarding the internal departmental requirements. Detailed layouts at 1/8″ and 1/4″ scales are developed to indicate medical equipment, and include locations of casework, outlets, medical gases, door sizes, and lighting requirements. The first interface with mechanical, electrical, and plumbing engineers occurs during early design development.

The final schematic design or preliminary design development drawings are often submitted to local and state code officials to solicit code and fire protection comments which must be accommodated in the plans to secure the final construction permit. At the conclusion of design development, the final hospital approval should be sought concerning partitions within and between departments. The final 1/8″ architectural background drawings based on design development are then submitted to the engineers who initiate calculations for the network of structural, heating, ventilating, and air conditioning systems to be detailed during the construction documents phase.

Construction documents are the final plans from which the building is built. Plans are drawn at 1/8″, 1/4″, and 1/2″ scales as particular areas may require certain levels of necessary detail. For example, a typical patient room or operating room would be developed at 1/2″ scale to show toilet accessories, medical headwall systems, and different medical gas features. During the construction documents phase, hardware, architectural finishes, and architectural specialties are indicated, including door handles and closures, carpeting, wall covering, tile, locker types, and internal glass.

The network of engineering systems is finalized in construction documents to include ductwork, conduit, shafts, vents, heatcoils, fans, cooling towers, etc. Final coordination among the various engineering disciplines and the architectural design occurs during the construction documents phase. Construction documents are also referred to as working drawings. For small uncomplicated projects the construction documents may be developed directly from schematic design.

Record drawings or as-built drawings are developed during the construction phase to reflect changes from the construction documents for field conditions, design errors and omissions, and scope changes.

Agency Reviews and Bidding

The review of design by state and local agencies and the bidding process may be concurrent activities. These activities must be closely coordinated, since code comments must be either reflected in the plans on which the contractors are bidding or in subsequent change orders during construction.

Agency Reviews

There are state building codes relative to life safety issues and fire protection which generally must be reflected in the construction plans before a building or construction permit will be issued. Often, there may be county or local municipal ordinances which govern construction projects and may require that a local building permit be processed prior to initiating construction.

The architect should advise the hospital on which codes apply to that institution's project, and the building design should incorporate code compliances which will be required to secure a building permit. The architect should start meeting with the government reviewing agencies in the mid to late schematic design stage to review the plans developed to date and to seek the agencies' reactions and comments. Generally, agencies will not commit in writing to variances or code interpretations prior to the final review of construction documents.

When the design is 90 to 95 percent complete, the plans should be submitted to the appropriate agencies for final review. Often it is required that a written functional program narrative be submitted as well. The agencies will then respond with written comments should changes be required. Typically, the construction permit will be granted contingent on revising the plans to reflect the required changes. This allows the hospital to proceed with construction although the design revisions may not be complete.

Design revisions to respond to review agency comments may be processed in three ways:

1. If the plans have not yet been distributed to contractors for bidding, the plan distribution may be deferred until the design changes are completed.
2. If the plans are ready for distribution or were already distributed for pricing, the bidding phase could proceed, and concurrently the design changes could be drawn as addenda to the original bid documents. In this case, the addenda are distributed to the contractors with reasonable advance time, before they submit their bids, so that their bids can reflect the design changes.

3. The design changes can be processed as change orders to the contract documents (documents upon which the construction contracts are awarded). In this situation, there may be a trade-off between schedule and cost, since this approach accelerates the construction start, but change orders may be more costly than competitive bids. Efforts should be made to minimize change orders once the contracts are awarded to avoid associated cost premiums.

Additional permits may be required from local agencies during the construction phase. For example, the mechanical subcontractor may be responsible to arrange for a review agency to inspect the mechanical systems upon completion of that portion of the work.

Bidding

Upon completion of design, the construction documents are distributed to interested contractors for bidding either concurrently with or after the agency review period. There could be a negotiated bid arrangement with a general contractor (GC), rather than a competitive bidding process. Under the negotiated arrangement, the GC provides construction cost and schedule information to the hospital during the design period and negotiates the final bid price with the hospital.

Contractors' bids are based upon narrative information and instructions and on the construction documents. The final drawings and narrative material, upon which the awarded contract prices are based, are together called contract documents. The narrative material includes:

1. Instructions to bidders regarding bid bond requirements, format for bidding, and applicable bid conditions
2. Finish schedule and outline specifications which indicate customized requirements, brands and types of material, and detailed descriptions for typical items or areas
3. Front ends which include the construction schedule, payment procedure, safety requirements, insurance requirements, and the form of the contract itself

A prebid conference may be held with the contractors, construction manager, architect, and owner to answer contractors' questions regarding the plans, to explain addenda, and to assure that the plans are correctly interpreted.

In a competitive bid situation, the bids received are analyzed and the contracts are generally awarded to the qualified low bidder(s). Under a negotiated bid arrangement, the GC's proposed price is reviewed by the architect and hospital and accepted or modified as appropriate.

FINANCING

Typically, the permanent long-term financing for a major capital construction program is not finalized until the design is complete and firm bids are in hand. Alternatively, financing may occur at the time when a guaranteed maximum price (GMP) is available, which could occur prior to completion of design. A GMP is a firm contractual price provided by a contractor who guarantees to build the building for a given price based on given plans and specifications. GMP is discussed in detail in the following section, as are the risks of an early GMP.

Firm bids or a GMP are generally desired prior to completing the permanent financing transaction, since the project budget should be well defined in order to accurately size the required debt proceeds. A funding shortfall could occur if the financing is based upon a preliminary cost estimate which is below the actual construction bids. Or, if the bids received are below budget subsequent to completing the financing, overborrowing would have occurred, which may not be prudent.

Preparation for the permanent financing should be done before the receipt of bids or the GMP, so that all financing activities are complete at the appropriate time and dovetail with the time when funds are required to start construction. The construction contract form should be finalized prior to financing. Should the financing be out of sequence and occur after construction is initiated, the hospital would have incurred a risk in signing the construction contract without having all funds in hand to complete the work. There are alternatives for structuring the financing schedule to meet the needs of a particular project. These alternatives are reviewed in a later chapter.

A traditional vehicle for long-term financings of major hospital capital expenditures has been publicly issued tax-exempt revenue bonds. Preparation for the permanent financing for a tax-exempt bond includes:

1. Finalizing assumptions for the financing plan
2. Developing legal documents, official statements, feasibility studies, reports, and certifications
3. Preparing a presentation for one or more of the national credit rating services to obtain a bond rating
4. Obtaining approval from the issuing public authority or municipality
5. Marketing the bonds
6. Signing the purchase offer contract with the investment banker which fixes the interest rate of the bonds
7. Closing the financing including signing all legal documents and

required certifications and receiving delivery of bond proceeds for investment by the bond trustee

Alternative vehicles for long-term financing include:

1. Tax-exempt private placements with a bank or insurance company
2. Direct taxable loan with bank or insurance company
3. Taxable public bond or note issues
4. Tax-exempt industrial development revenue bonds
5. Federal Housing Administration (FHA-242) mortgage loans, in possible combination with loan guarantee by the Government National Mortgage Association (GNMA)

Interim or short-term financing may be appropriate for some construction programs. Options include:

— Line of credit
— Public or private tax-exempt notes in possible combination with letter of credit security
— Bond anticipation notes
— Short-term tender bonds
— Direct construction loan with bank or insurance company

Many state financing authorities have a short-term tax-exempt financing program for financing major capital equipment from a pool of outstanding note proceeds at an established interest rate.

A more detailed discussion of the financing vehicles is to be found in Chapter IX.

Construction

The type of contractual arrangement which the hospital has with the contractor will determine the construction approach to be used. The most frequent construction approaches are traditional general contractor, fast track, phased construction, design build, and construction management.

Traditional General Contractor

The hospital engages an independent architect (and possibly independent engineers and consultants) who prepares the *complete* construction package consisting of complete plans, including architectural, structural, plumbing, mechanical and electrical drawings, and complete and detailed specifications. A general contractor is then engaged to construct the project as designed by the architect. The construction contract may take one of the following forms:

1. *Lump sum*—the construction package is submitted to several contractors who would each submit a lump sum, fixed price for the construction of the project. The contract is then awarded to the low bidder.
2. *Guaranteed maximum price* (GMP)—a variation of the lump sum approach, the contractor submits a "not to exceed" price for the project with the possibility that the project may, in fact, cost less. The contract itself will define the circumstances under which the price may be less which are usually based upon the cost of materials and labor purchased (e.g., subcontracted) by the contractor.
3. *Cost plus*—a contract may provide for completion of the project on a "cost plus" basis, i.e., cost plus a fixed fee or percentage fee.
4. *Negotiated contract*—this approach can be utilized with either a lump sum or GMP contract. The difference is that rather than putting the construction package out for competitive bids, a contractor is selected for the project. This approach is used when a particular contractor is desired because of either its favorable reputation or its unique ability to construct the project.

Fast Track

The primary difference between the traditional approach and the "fast track" approach is that incomplete plans and specifications (basic documents, schematics, preliminary structural drawings, etc.) are utilized in contracting with the general contractor. The advantage of this approach is that the length of time from the inception of the project to its completion can be reduced since the design and construction phases are overlapped. The incomplete construction package is usually submitted to several contractors for preliminary estimates or proposals. One of the contractors is then selected for further negotiations. The construction contract may take one of the following forms:

1. *Lump sum*—the construction contract may provide for the construction of the project at a lump sum price. The danger with this approach is that (a) the contractor will necessarily include a potentially large contingency in his proposal because of the incomplete plans and specifications or (b) the contract will have a very narrowly designed scope such that changes resulting from refine-

ment and detailing of the incomplete plans and specifications will result in increases in the lump sum proposals. The use of a lump sum contract in connection with fast track construction increases the risks inherent in the lump sum proposal.

2. *Guaranteed maximum price* (GMP)—the construction contract may provide for the construction of the project at a "not to exceed" price. The GMP will, as in the case of the lump sum proposal, necessarily include a potentially large contingency. Frequently, a GMP arrangement will include allowances for items not specified at the time the proposal is given. Allowances are undesirable and should be avoided whenever possible.
3. *Cost plus*—the construction contract may provide for the construction of the project at cost plus a fixed or percentage fee.
4. *Cost plus with guaranteed maximum price*—the construction contract may provide that the contractor will begin the project on a cost plus basis, as limited by an established construction budget, with a GMP to be established once the plans and specifications are complete.

The risks inherent in fast track construction are: (1) Because all engineering coordination is not completed at the time that the contract for the early phases of the work is awarded, change orders during construction may be required to achieve the proper coordination; and (2) The potential for litigation is increased with respect to whether the completed plans and specifications represent a change in scope from the preliminary drawings (and, therefore, an increase in cost) or represent completion and refinement of the preliminary drawings (and, therefore, no increase in cost). Hence, fast track construction requires a certain degree of sophistication and experience on the part of the owner.

Phased Construction

The construction work in this approach is bid in phases. This may be according to the fast track construction approach whereby the design is completed and bid in phases. In that case, construction is initiated prior to completion of all design. Alternatively, upon completion of all design, it may be prudent to bid and construct the work in phases rather than awarding a lump sum construction contract or multiple subcontracts at one time. For example, bidding the renovation phase of a project may be deferred until nearer the actual start of work, since renovation is often the final phase of a new construction and remodeling program. The various contracts for phased construction would take one of the forms described previously for the traditional general contractor.

Design Build

The owner contracts with a single entity for both the design and construction of the project. The "contractor" may be a joint venture consisting of an architect, an engineer, and a contractor or it may be a contractor who personally engages an architect. The economics of the construction contract would take one of the forms described previously for fast track.

Construction Management

Under this arrangement, a single entity, acting as the hospital's "agent," administers the work of various subcontractors. Typically, all work for the project is performed by subcontractors except for the general condition items (job site trailer, office equipment, fences, barricades, temporary facilities) provided by the construction manager. The contracts of the various subcontractors may be executed by the construction manager or the owner, depending on the agreement between the owner and the construction manager.

The construction management arrangement provides the owner with the experience and expertise which it would require to function as its own "general contractor" for the project. The construction manager, at least theoretically, utilizes his/her experience and expertise solely for the benefit of the owner in negotiating with subcontractors and coordinating construction. As a general contractor, the owner receives the benefit of the low bids obtainable for each trade category of the work. By comparison, under traditional arrangements, various subcontractors and a single general contractor may not have received the lowest bid for each work category received by all of the general contractors. The economics of the construction management agreement could take one of the following forms:

1. *Cost plus*—the construction management agreement could provide for the construction of the project for the cost of the various subcontracts plus the cost of general condition items plus a fixed or percentage fee.
2. *Guaranteed maximum price*—The construction management agreement could provide for the construction of the project for the cost of the various subcontractors, plus the cost of general condition items plus a fixed or percentage fee, but in no case more than a guaranteed maximum amount.

The foregoing overview of major approaches to the construction phase is intended to describe the normal relationships between the owner and the construction manager or contractor. The exact nature of the relationship in any given instance will be governed by the agreement between the owner and the contracting party. Therefore, the importance of the agreement itself cannot be overemphasized.

While the preceding provides an overview of major approaches to the construction phase, Chapter X addresses in detail such construction issues as shop drawings, change orders, warranties, etc.

OCCUPANCY AND STARTUP

Upon completion of construction, the occupancy and startup period commences. This period may take several months to complete, depending on the size and nature of the project. It is essential to recognize the differences between substantial completion of construction, final completion of construction, and beneficial occupancy. Although an architect may certify that a project is substantially complete, final completion may be several weeks or even months off while mechanical and electrical systems are tested, warranties are processed, and minor construction problems are remedied. During this period it may also be necessary for state and/or local agencies to conduct a walkthrough in order to approve and issue a certificate of occupancy, depending on the nature of the project and the type of construction.

The hospital should plan an appropriate time period for cleanup and employee orientation subsequent to substantial completion but before beneficial occupancy. Thus, for budgeting purposes, it should not be assumed that the construction completion date is the same date that revenue production of new facilities will occur. The construction contract could provide for temporary labor to orient the employees to the new systems.

Should construction be completed in distinct phases, the hospital should plan ahead to occupy the completed facilities on a phased schedule. This allows for the expensing of interest, if any, and depreciation in a timely way, thereby enhancing the hospital's cash flow. The construction contract should provide for phased occupancy, if applicable, and the avoidance of dust and noise. A provision barring claims for interference during phased occupancy should be stipulated in the contract.

BAR CHART SCHEDULE

The bar chart schedule indicates the various project phases and major work tasks required to implement a capital construction project. The following chart is not meant to suggest a prescribed time frame, since each project has its own schedule requirements. Rather, it is meant to illustrate a general scheduling approach and the time relationships of the various work tasks.

Implementation of a Capital Construction Project (Bar Chart Schedule)

Work Tasks	Months																																		
	1	2	3	4	5	6	7	8	9	10	11	12	13	14	15	16	17	18	19	20	21	22	23	24	25	26	27	28	29	30	. . . . .	55	56	57	58
1. Completed work tasks before the project is initiated																																			
A. Strategic plan	X																																		
B. Preliminary program development	X																																		
C. Long-range financial plan/ debt capacity analysis	X																																		
D. Master physical facilities planning	X																																		
2. Appoint project committee	X																																		
3. Project team selection		X																																	
4. Project team meetings			X	X	X	X	X	X	X	X	X	X	X	X	X	X	X	X	X	X	X	X	X	X	X	X	X	X	X	X	. . . . .	X	X	X	X
5. Project definition																																			
A. Establish need			X	---	X																														
B. Test demand			X	---	X																														
C. Define scope				X	---	---	X																												
D. Prepare master schedule				X	X																														
E. Develop budget							X	X																											
F. Prepare financing plan								X	X																										
G. Test debt capacity									X	X																									
6. Governing board approve in concept										X																									
7. CON phase																																			
A. Prepare application					X	---	---	---	---	---	X																								
B. Hearings and approvals												X	---	X																					
8. Design																																			
A. Block drawings					X	---	X																												
B. Award A/E contract														X																					
C. Schematic design														X	---	X																			
D. Design development																	X	---	---	X															
E. Construction documents																					X	---	---	X											
F. Record drawings																																X	---	---	X
9. Agency reviews																X				X			X	---	X										
10. Bidding																									X	X									
11. Financing																																			
A. Documentation																						X	---	---	---	X									
B. Rating																										X									
C. Governing board approval																											X								
D. Pricing and commitment																										X	X								
E. Closing																												X							
12. Construction																																			
A. Award contracts																												X							
B. Construction																												X	---	---	. . . . .	---	X		
C. Test systems																															X	---	---	X	
13. Occupancy and startup																																	X	---	X

IV

The Project Team

This chapter addresses the role and function of the participants who will develop the project, including the responsibilities of hospital administration. The manner in which professional disciplines are selected, organized, and managed is reviewed and general comments regarding professional contracts and agreements are included. Finally, a suggested vehicle for direct involvement of the governing board and the medical staff in the work of the project team is discussed.

It is important to recognize that the planning and implementation of a major construction project may require assistance from a variety of outside experts, entailing a significant financial investment by the hospital. While the hospital staff may be competent in certain areas, other tasks may be highly technical and are best delegated to an appropriate specialist. It is the chief executive officer's (CEO's) responsibility to determine which functions could be accomplished by in-house personnel and which tasks are better provided by consultants. This decision will depend on the size and nature of the proposed project and the depth of the hospital staff and its other obligations. As the hospital staff is responsible for the day-to-day operations of the institution, their ongoing work loads may not allow much additional time for project planning and implementation activities. Therefore, either outside assistance or additional hospital personnel may be required to meet the extensive time commitment of the project planning and implementation process.

It is also important to recognize that hospital administrators are trained to operate health care institutions and are not by trade functional designers, statisticians, or capital debt specialists. Thus, even given time availability, hospital personnel may not be the appropriate parties to assume certain project functions. The most efficient use of time and dollars will result by assigning project responsibilities to the most experienced professionals, be they internal or external to the hospital.

Project Team Players

Since planning and implementing a major construction project is a multifaceted interdisciplinary process, the project team will necessarily be composed of players representing many disciplines. Some of the team disciplines may overlap and others may be combined under a single representative, e.g., architectural services may include engineering, interior design, and equipment planning. Certain members of hospital administration must consider themselves members of the project team and act as such by keeping commitments to attend meetings and completing appointed tasks as outside consultants are expected to do. The following subsections address the specific roles and responsibilities of possible project team players, although all functions may not be appropriate for every project.

Hospital Administration

The project team players from the administrative staff principally include the following:

- Chief executive officer
- Chief operating officer
- Chief financial officer
- Director of planning
- Director of plant operations, plant engineer, or similar position

In addition, during specific phases of the project development, other hospital staff will be closely involved with the planning activities. For example, during the program phase, the associate administrator(s) should be closely involved with the planning; and during the schematic design and design development stages, nursing administration and managers of departments affected by the project will be meeting with the architects to review design issues. The following are suggested assignments of project responsibilities among hospital administration. However, the responsibilities of the administrative staff could vary with the size and nature of the decision and project.

The CEO's responsibilities relative to the project could include:

1. Assuring that the project is consistent with the mission, philosophy, role, goals, and objectives as specified in the institution's strategic plan and program development statements

2. Assuring that all necessary governance and regulatory approvals are secured
3. Integrating the input of the planning and finance functions
4. Assuring appropriate input to the project from the medical staff
5. Assuring appropriate and consistent interaction with public and community bodies, including Health Systems Agencies, state health agencies (covering 1122/CON, building, and code issues), state and local fire marshalls, county agencies, media, fund raising groups, Blue Cross, etc.

Many of the above responsibilities could be delegated to others for implementation, but because of the number of often competing aspects of the project development process, the final authority for assuring that these responsibilities are met must rest with the CEO. The CEO may wish to delegate responsibility for the day-to-day project development to the chief operating officer.

The chief financial officer's project responsibilities could include:

1. Preparing or reviewing and approving all key budgetary and financial data developed in connection with the project, including debt capacity studies, CON budgets, financing plans, projected financial statements, financial feasibility studies and reports, project cost expenditure reports, construction contracts and agreements, construction change order logs, and reports of the bond trustee
2. Assuring that the project meets the financial guidelines established by the institution
3. Interacting appropriately with the financing team regarding the development of financing and implementation plans of any long-term financing transaction, including preparing the official statement, developing the financing schedule, developing the feasibility study, and preparing for the rating process
4. Reviewing and negotiating all legal and financing documents associated with any debt transaction and advising the CEO regarding the effect of any significant business covenants on hospital operations
5. Advising the project team on the financing schedule, capitalization, and reimbursement issues
6. Monitoring and reporting on project cost expenditures relative to the budget

The project responsibilities of the director of planning could include:

1. Advising the CEO on the conceptual project planning as it relates to the hospital's long-range plan and program development
2. Interacting with the facilities planner relative to historical and projected utilization statistics, as required for the space plan
3. Preparing or reviewing and approving all materials developed for the CON process and being responsible for interaction with local and state health planning agencies
4. Monitoring the design process to assure that the design properly reflects the programmatic intent of the project

The project responsibilities of the director of plant operations or plant engineer could include:

1. Advising the project team relative to the condition of existing facilities and plant engineering issues
2. Participating in design work sessions and reviewing and advising on all design plans, particularly on construction documents
3. Reviewing all budgets for construction costs and fixed equipment
4. Interacting with the construction manager and architect during the construction and occupancy periods to assure correct operation of the plant and systems

The above listings are not meant to be inclusive and there may be shared responsibilities among staff. The following paragraphs address additional items of note relative to the in-house staffing structure in support of the project.

The "owner's representative" is a technical term defining the position of authority for the day-to-day construction phase. This individual is generally charged with approving the contractor's pay requests (after the architect's approval) and change orders, and with interacting with the on-site construction management staff or the general contractor. He or she may or may not be the director of plant operations depending upon the size of the hospital and its organizational structure.

Some institutions may have a full-time person on staff to administer construction projects, and he or she could also act as the owner's representative during the construction phase. The construction administrator position may also be responsible for directing the day-to-day activities of the design process. In other hospitals, it may be more appropriate for the director of planning or an associate administrator to direct the design team, and then during the construction phase, this responsibility might be transferred to the plant engineer or a staff construction administrator.

In the case of transferring project authority between project phases, a certain amount of discontinuity could occur. Then, the CEO's task to integrate the design and construction phases becomes more difficult, and accountability relationships could be blurred. It is essential to recognize, however, that during the planning and design phases, as with the construction phase, there must be an owner's representative or project representative who is charged with the day-to-day direction of the project activities, to serve as liaison between the design team, hospital administration, and department managers during program planning and design development.

The Project Manager

The concept of a project manager is different from the concepts of a project representative or an owner's representative. The project manager should be responsible for the day-to-day progress of the project, subject to the direction and supervision of the CEO and the governing board. The project manager should report directly to the CEO or another administrator with senior management authority for the project. There are hospital consulting firms which specialize in project management. The project manager may or may not be a hospital staff member, or, in the case of small simple projects, there may be no project manager at all.

A project consists of many phases: programming, regulation, design, financing, and construction; and there is much cross fertilization between different areas of expertise during the project development process. Therefore, the project manager who is knowledgeable regarding the integration of all project phases and disciplines will be more successful in implementing the project than would a project manager whose knowledge is confined to a particular phase or discipline. The project manager should be responsible for the project from the inception of planning through the completion of the construction. The responsibilities of the project manager could include:

1. Monitoring the work of the project team relative to task accomplishment
2. Developing and monitoring the master project schedule and integrating and dovetailing the work of the project team into the schedule
3. Monitoring the project budget
4. Developing and maintaining a system for communications among the project team
5. Maintaining a master project file containing records or other documentation regarding all key project elements, including budget, schedule, financing plan, project team meeting minutes, etc.

6. Developing project checklists of all pending issues for attention by the project team
7. Assuring consistent integration of the work of the project team relative to CON applications, financing documentation, reports for governance, and other interdisciplinary work products
8. Advising the CEO relative to project crisis management and all important project issues
9. Integrating observations of all disciplines into comprehensive project reports as required
10. Identifying project problems and recommending appropriate solutions

When no one party has project management responsibility, the CEO must integrate the recommendations and reports of the various parties having partial responsibilities. In this case, the CEO is essentially assuming responsibility for the project management. This should be a conscious decision. The CEO must recognize the extent of time and effort necessary to properly manage the project. Generally, a CEO's overall hospital management responsibilities do not permit the amount of time required for full-scope project management. The decision to include a project manager on the project team should be based upon the size and nature of the project, the additional time commitments that hospital staff can make, and the experience and knowledge of hospital staff relative to the various stages of the project development process.

The Functional Planner

There are hospital consulting firms who specialize in facilities and functional planning. The functional planner engaged for the project may be the same firm who assisted the hospital in developing its long-range plan, role, and program study or master facilities plan. The functional planner's role typically includes:

1. Assisting in program development activities
2. Translating projected volume and workloads into space requirements
3. Reviewing existing facilities and advising on life safety code violations, systems operations, functional flows within and between departments, and staffing patterns
4. Preparing a facilities program narrative for all hospital departments and areas to include location, relationships, operational needs, design criteria, and room elements

5. Advising the hospital and architect during the design process regarding departmental relationships and layouts, internal traffic patterns, and fixed medical equipment
6. Assisting the hospital staff during the building startup period regarding operational systems

The functional planner may have consulting capabilities regarding planning issues and therefore may also assist in developing the CON application.

The Architect and Engineer

Generally the architect and engineer (A/E) are grouped as one discipline since the work of each is inextricably related in the development of the design. Some full service architectural firms have engineering capabilities while other architects are strictly design firms which associate with independent engineering firms. The main engineering disciplines involved in a hospital construction project include mechanical, electrical, plumbing, civil, and structural, and these disciplines may be represented by more than one firm on a particular project.

During the project development, the senior hospital administration will relate more closely with the architect than with the engineer, since the architect usually manages the engineering design, except for engineering projects, such as a boiler plant expansion.

The A/E's role typically includes:

1. Assisting in master physical facilities planning activities
2. Translating the facilities program narrative and space projections to block drawings
3. Proceeding with design through schematic design, design development, and construction documents
4. Assisting in securing building permits, including meeting with code officials and the fire marshall
5. Reviewing all narrative bid specifications prepared by the construction manager (CM), or if there is no CM, preparing all bid specifications
6. Assisting in the bidding process and reviewing and advising on all bids
7. Observing the construction process to determine if the work is proceeding in accordance with plans and specifications and issuing certificates required in connection with financing
8. Reviewing and approving shop drawings and samples during construction

9. Preparing change order documentation and approving change orders during construction
10. Reviewing and approving contractor's applications for payment
11. Conducting final building inspection to determine substantial completion

The A/E should have hospital experience. The architect may also be engaged to provide landscape design, medical equipment planning, and interior design.

The Construction Manager/Cost Estimator/ General Contractor

The construction expertise required to plan and implement a capital expansion project may be provided by one of several types of firms, depending on the nature of the project. A construction manager or cost estimator may be retained early in the planning process to provide cost projections. The construction manager could also manage the actual construction process. Alternatively, the general contracting approach could be utilized whereby the low bid general contractor would manage the construction. The hospital's decision to use either a construction manager or a general contractor should be based upon the size and scope of the project and specific preferences for the particular approach under either construction management or general contracting. The following subsections explain these types of firms in greater detail.

The Construction Manager. Construction management (CM) grew out of the need to have valid information regarding construction costs and scheduling prior to the construction period, due to cost overruns and delays on many projects, particularly in the government sector. With the advent of Public Law 92-603 and Public Law 93-641, the regulations concerning reimbursement for capital expenditures forced hospitals to develop accurate project cost projections early in the design process. A number of national construction firms began to specialize in hospital CM to meet the need for construction expertise and consultation during the preconstruction phase.

Most CM firms originated as general contracting firms. Most major hospital CM's continue to practice general contracting as well and therefore have current cost data and continuing construction expertise. This type of firm would generally manage the construction of the project as well as provide preconstruction consultation. Some firms provide preconstruction consultation only, they do not manage the construction process, and are usually architecturally oriented.

The role of full service CM includes:

1. Providing construction cost estimates based upon each major phase of design, including program phase, schematic design, design development, and construction documents
2. Providing value engineering services relative to potential cost reductions through use of alternative systems and materials
3. Estimating the construction schedule and advising on construction phasing
4. Reviewing all design plans and specifications relative to coordination of construction trades, omissions, and duplications
5. Managing the bidding process including preparation of instructions to bidders, prequalification of bidders, conduct prebid conferences, analysis of bids, review of bid bonds, interview low bidders to confirm their understanding of the work, and recommendation on award of construction contracts
6. Preparing and administering construction contracts
7. Preparing a critical path method schedule for all trades
8. Acting as the owner's agent in the management and supervision of all construction work
9. Supervising the processing and approval of shop drawings and samples
10. Determining the adequacy of contractors' personnel, equipment, and schedules for availability of materials
11. Coordinating the owner, architect, and contractors relative to the interpretation of plans and specifications
12. Maintaining job site records including shop drawings, contracts, cost accounting records, daily log of job site activities, schedules, and as-built drawings
13. Reviewing and advising on change orders and administering their processing

The CM may not perform any of the construction work itself, but may provide general conditions items, including cranes, hoists, barricades, fencing, temporary toilets, and temporary signage. There are many forms of construction management and contracting: the owner/agent approach, guaranteed maximum price (GMP), fast tracking, and negotiated bid. These are all addressed in detail below.

The Cost Estimator. In cases where a construction manager is not engaged to provide preconstruction consultation, a cost estimator may be necessary to provide early and accurate construction cost consultation. This may be deemed a requisite for CON filings or financial planning purposes. Many architectural firms subcontract construction cost estimating to specialized firms as a check during the design process.

The General Contractor. A general contractor (GC) may be utilized to construct the project in lieu of the construction management method. In such a case, a cost estimator may be engaged for early cost analysis. The GC typically is selected based upon the single low bid for all construction work, and is therefore not selected until the design is complete. (See Chapter III on the various construction approaches.) The GC contracts directly with the subcontractors and manages their work. Typically, the GC also performs the work of certain construction trades, i.e., excavation and sitework.

Under a GC arrangement, the architect is required to provide more extensive services relative to the bidding, bid negotiation, and contract preparation and award. The GC does not act as the owner's agent as in the CM relationship. The degree to which construction records are open and the owner represented in the day-to-day construction could be more limited under a GC arrangement than with a CM.

The Design/Build Firm

The preceding subsections defined the traditional design and construction team as an architect and an engineer together with either a construction manager or a general contractor, with possible input from a cost estimator. An alternative to this approach is to engage a design/build firm which would be responsible for both the design and construction phases. This type of firm would have all architectural, engineering, and contracting expertise in-house.

One possible benefit of the design/build approach is that the hospital has fewer different consultants to coordinate. Combining the design and construction disciplines, however, may remove some of the healthy tension and challenge that occurs between these two areas of expertise, and that often produces a better solution. The design/build approach may be appropriate for projects involving little custom design and minimal unique owner requirements, such as medical office buildings or parking garages.

The Financial Advisor

In some instances it is appropriate to engage an individual or firm with specialized expertise to assist the chief financial officer with certain financial aspects of the project. The financial advisor should be in a strict advocacy position on behalf of the hospital, as opposed to the feasibility consultant or investment banker, who may have fiduciary relationships with others, as discussed in the following subsections.

The financial advisor should be engaged early in the process to prepare a strategic financial plan or debt capacity analysis. The financial advisor's role typically includes:

1. Analyzing debt capacity
2. Evaluating construction budgets and preparing overall project budgets, including professional fees, administrative expenses, equipment, and contingencies
3. Preparing projected financial statements for the CON application
4. Monitoring and revising as necessary the project budget
5. Advising on assumptions for the plan of financing
6. Preparing alternative financing plans and advising on alternative financing vehicles
7. Advising on phasing of capitalized interest and debt amortization options
8. Preparing the financing schedule based upon the requirements of the design and construction schedules
9. Structuring the plan for use of hospital equity and monitoring the actual equity disposition
10. Preparing cash flow projections and monitoring the budgeted to actual cash flow
11. Monitoring and reporting on project cost expenditures
12. Coordinating the work of the financing team
13. Negotiating business covenants in the loan agreements
14. Preparing the hospital section of the official statement
15. Preparing plans for interim financing
16. Preparing the rating agency presentations and booklets
17. Monitoring and reporting of income earned on invested funds during construction

The Financial Feasibility Consultant

The financial feasibility consultant is generally a public accounting firm, although some management consulting firms provide financial feasibility services. "Financial feasibility study" is a technical term applying to independently forecasted financial statements which indicate the hospital's ability to service its debt including the debt related to the project.

For most publicly issued tax-exempt bond transactions, a popular financing source for hospital projects, a financial feasibility study is required to market the debt. The hospital may engage a feasibility consultant to prepare a preliminary feasibility study or debt capacity analysis early in the planning process. The role of the consultant typically includes:

1. Assessing historical utilization and current trends in use rate in order to forecast future use rates
2. Assessing historical and projected population trends
3. Assessing the hospital's market share and capital plans of competitive institutions
4. Projecting future utilization based upon the use rate, market share, service area, and medical staff
5. Assessing the rationale and assumptions for changes in revenue and expenses
6. Projecting future operating expenses
7. Projecting financial statements based upon the projected utilization, projected expenses, the project, and the proposed debt

The Investment Banker

The investment banker typically structures the mechanics of the debt transaction, interacts with the finance authority and the rating agencies, advises on the legal financing documents, and markets the debt. In developing preliminary financing plans, the hospital should interact with the investment banker in structuring the plan of financing relative to current market rates and trends in the capital market place.

Legal Counsel

Legal counsel should advise on all legal matters related to a particular project. Often, a special legal counsel may be engaged for specific tasks or issues. The role of legal counsel in the capital construction process typically includes:

1. Reviewing and advising on all professional contracts
2. Advising on CON legal issues, hearings, appeals, and negotiations
3. Advising on zoning issues
4. Developing, reviewing, and advising on legal documents associated with financing transactions; and
5. Rendering an opinion as to the legality of financial documents and transactions

For long-term debt transactions, the role of legal counsel is significant. This is discussed in detail in Chapter IX.

The Fund-Raiser

Many hospitals may already have an ongoing long-term development program, which is an excellent vehicle to build equity for capital expansion projects. In addition, the possibility of conducting a fund-raising campaign to help support a specific building project should be considered for major capital construction programs. The hospital may wish to engage a professional firm to evaluate the feasibility of a campaign and/or to direct the campaign activities. A fund raising feasibility study is generally based on approximately 60 to 90 interviews with hospital administration, board, and medical staff as well as leaders of the community, business, and industry. The interviewees are asked to comment on the image of the hospital, capabilities of its staff and physicians, and the individual's giving priority to the institution. Based upon the study, the consultant will indicate if a campaign appears to be feasible and will project a goal for the campaign.

Should a campaign be initiated, the fund-raiser's role could include:

1. Assisting in recruiting and orienting the campaign volunteer leader and workers who will conduct the actual solicitation
2. Developing public relations materials including brochure, case statement, and audio visual presentation
3. Developing press releases and coordinating media events
4. Planning and coordinating campaign kickoff activities, volunteer dinners and meetings, and any solicitation events and programs
5. Creating pledge cards and maintaining records of pledges and cash received
6. Monitoring and motivating the work of the volunteers
7. Estimating projections of cash proceeds for specified future periods
8. Assisting the hospital to establish a permanent development office

Generally a professional fund-raiser does not directly solicit donations. The role of the fund-raiser is to organize the campaign and volunteer recruitment program, then to coordinate and monitor the follow-through of the volunteer solicitation program. Board members must understand that their time commitment to a capital campaign may be significant and that their active involvement in volunteer recruitment and actual solicitation will be essential to assure a successful campaign.

Other Project Team Members

Depending on the needs of the project, other team members who could be required include separate CON consultants, an equipment planner, a landscape architect, an interior designer, a code consultant, a parking consultant, and consultants for specific clinical programs, e.g., radiation therapy. The hospital should determine whether other team players are required based upon discussions with existing team members and a review of their capabilities, capabilities of hospital staff, and the special needs of the project.

Project Team Selection and Organization

Once a decision is made to engage certain professional disciplines for the project, e.g., an architect or a construction manager, the selection process should be deliberate, structured, and rigorous. Should the hospital be uncertain regarding the applicability of a given professional discipline to the project, the selection process could be structured as a potential two-step education and selection process.

The selection process should include the following steps:

1. Assemble the selection committee, a small group consisting of key administrative staff members, governance, and medical staff representatives (this could be the project committee as defined in a subsequent section)
2. Prepare request for proposals (RFP) delineating project scope, outline of work, proposal format, due date for proposal, and interview date
3. Compile a list of potential recipients of RFP including nationally recognized firms and firms with successful experience on other hospital projects
4. Prequalify candidates against established criteria and mail RFP to prequalified firms
5. Prepare a list of selection criteria for the selection committee
6. Receive proposals, review for completeness, and prepare written summaries
7. Review proposal summary and selection criteria and reduce the list of candidates as appropriate
8. Interview a short list of candidates and select the finalist

It is helpful to include one or two medical staff representatives on the selection committee since involvement in project-related activities could educate the physicians regarding the project, stimulate physician enthusiasm, and generate their support for the project. Preparing a comprehensive RFP is important to the selection process since it provides the basis for the eventual fee proposals. The RFP should therefore clearly define the scope of the engagement and provide a detailed format for the proposals, which will also assure uniformity in comparing the proposals. Firms with a national reputation in their particular field should be included in the list of candidates since the specialty nature of hospital capital projects requires experienced technical expertise and since a national firm could provide a flavor of industry trends in the particular field. Firms which specialize in health care may have experience with problems similar to those the hospital's project entails. Local firms should demonstrate the same qualifications as any experienced specialized firm which may not be local.

Subsequent to selecting the professionals for the project team, the entire team must be organized with a clear understanding of accountability and reporting relationships. These relationships are addressed in detail in the sections Project Team Players, Contracts, and Managing the Project Team. Generally, a good project team organization requires:

1. A written outline, statement, or similar documentation regarding the specific services or tasks to be performed by each party
2. A master project schedule and task list which integrates the milestone activities of all team players
3. Commitment to hold regular project team meetings
4. A system for communication among the team and for monitoring project decisions, work tasks, schedules, and budgets

CONTRACTS AND AGREEMENTS

Although professional contracts and agreements are reviewed by the legal counsel, this task should not be completely delegated to attorneys with no hospital input. The hospital should be familiar with the terms of all professional contracts to assure that the needs of the project are met, that no unnecessary overlap or duplication of services occur between firms, that no conflicts exist between contracts and that all required services are covered. This is particularly important relative to the contracts with the architect and the construction manager.

It is important that the hospital take a negotiating posture relative to professional contracts. Clauses which are not clear should be challenged and explained. Standard contract forms prepared by a particular discipline (e.g., American Institute of Architects, Associated General Contractors) are often biased in favor of that discipline and are always negotiable.

It is also important that contracts be negotiated early in the process and certainly prior to commencing work to avoid subsequent disputes regarding compensation for completed work. Should specific circumstances or the scope of services not be finalized when negotiating the contract, i.e., need for a guaranteed maximum price for construction, then the contract should be developed with options or converter clauses to provide for possible alternatives.

When the terms of a contract provide for compensation based upon time and/or materials, a fixed cap should be negotiated, and the owner's approval should be required prior to proceeding past that cap. When certain expenses are to be reimbursable under the terms of the contract, either a cap should be negotiated or an estimate provided, and again, the owner's approval should be required prior to exceeding the estimate.

Managing the Project Team

This section addresses general techniques for managing the work of the project team. The group of professional disciplines engaged in the project development process, including consultants and hospital staff, must consider themselves team players. This point is important and should be emphasized to the team at the outset of the process. Commitment to teamwork will hopefully minimize secret agendas, working at cross purposes, and other inefficient and disruptive attitudes and activities.

The following is a suggested list of management techniques applicable to the team process (and which is not necessarily limited to the development of a capital construction project):

1. Regular monthly meetings of the entire project team should be scheduled; establishing a standing meeting day will avoid conflicts developing in busy calendars
2. An individual or a firm knowledgeable of all project disciplines should prepare the monthly meeting agendas, chair the meetings, and record and distribute meeting minutes and attendance lists; meeting agendas should be distributed to the team in advance and should delineate the various reporting requirements
3. At each monthly meeting, the status of the budget and schedule should be reviewed and revisions should be disseminated promptly
4. Tasklists should be developed and maintained and completion dates should be indicated

5. Between monthly meetings, there should be follow-ups with team members relative to task accomplishment
6. An individual or a firm should be designated as the contact point for team members to report problems and crises; team members should be directed to adhere to this reporting mechanism on an immediate basis
7. When problems are identified relative to the project, assignments should be made for developing and quantifying aternative solutions for decision by the hospital and/or the team

THE PROJECT COMMITTEE

In addition to the technical involvement of the project team, the project development process also should include input and direction from the hospitals' governing board and medical staff. Lacking such involvement, the project planning could proceed in a direction contrary to the desires and intentions of the governance and the medical staff, and could result in wasted time and money should revisions and backtracking be necessary. Medical staff involvement in the project development process is important—to receive input during planning and design, to educate the physicians to the project, and to stimulate physician support for the project.

The problem with confining governance involvement in the construction project to reports at the board committee and full governing board level is that:

1. Committees are segregated by such functions as planning, finance, and building while the project activities pertain to all such functions, and governance direction of the project requires a body integrating these functions
2. Meetings of standing board committees and the full board occur infrequently (probably monthly or less often), and numerous issues must be addressed at those meetings; the project may require more frequent and extensive discussion and direction

A suggested vehicle for direct involvement of the governing board and the medical staff is the formation of a project committee. The committee should be limited to a workable size, perhaps five to seven members, and could include two or three board members, one or two physicians, as well as key administrative staff.

The project committee should meet monthly following the project team meetings to receive the report of the team and act on project business as required. Hospital administration and the project manager should report on the work of the project team to the committee.

The committee would then be responsible to report on the progress of the project to the full board. The board may designate certain governance responsibilities to the project committee, including awarding professional and construction contracts and approving scope, budget, and schedule adjustments within the total parameters approved by the full board. In this case, the board would ratify the project committee's actions.

The role of the project committee is to act on behalf of the governing board within the bounds of the approved project scope and cost, to provide project direction, supervision, and approval in a number of areas. Specific charges to the project committee could include:

1. Award, modify, and change construction contracts
2. Approve, modify, and change consultant contracts
3. Receive budget and financial reports from the project team on a periodic basis, including recommendations for the reallocation of funds in the project budget
4. Provide direction to the project team on matters of overall project direction and construction progress
5. Review information from and provide advice to hospital administration regarding ongoing project construction implementation
6. Advise the governing board on the financial status of the project and formulate recommendations when necessary for changes in scope and/or cost of the project
7. Provide a postconstruction evaluation of effectiveness and completeness of the process to the governing board.

V

Program and Project Definition

The preceding chapters addressed master planning tasks needed prior to initiating a project, the composition and function of the project team, and an overview of the various project phases. The remainder of this guide discusses in greater detail the specific project phases.

This chapter addresses the key elements of the program and project definition phase, which include need and rationale, testing demand, scope, budget, financing plan, feasibility, and schedule. As indicated earlier, there is much overlap between this phase and the CON application preparation process. This section should be cross-referenced with the section on CON process.

Need and Rationale

The following are a number of factors upon which the need and rationale for a project could be predicated:

- To meet additional space requirements
- To improve functional configurations
- To provide facilities for additional volumes of service and/or new or expanded programs
- To correct code deficiencies or citations from the Joint Commission on Accreditation of Hospitals (JCAH) or the American Osteopathic Hospital Association (AOHA)
- To enhance productivity and promote operating efficiencies
- To improve the efficiency of energy systems

Although the above planning objectives may have been identified and evaluated in the long-range plan, they should be revisited and closely scrutinized relative to planning a specific construction project. Identifying the need and rationale for a project should always precede the definition of the project itself, since alternative approaches for meeting that need will be part of the project scope and definition.

During the evaluation of the need for the project, it would be helpful for the hospital to establish priorities of departments and services to be addressed in the building project. Priorities should be established on the basis of operating needs, cost, competitive considerations, medical staff desires, as well as the overall goals and objectives of the institution. Testing the demand and financial feasibility may result in constraints which limit the ability of the hospital to meet all needs through one specific construction project. It may then be necessary to reduce the proposed project scope based upon the established priorities.

Testing the demand for future services, which is addressed in the following subsection, should be accomplished simultaneously with evaluating the need and rationale for the project, since projections of future utilization will affect such need issues as space requirements, addition or expansion of programs, and operating efficiencies. Thus individuals from the planning, finance, and design disciplines must interact closely in completing their early work tasks.

During the early code analysis as discussed in Chapter II, the hospital should have compiled a list of violations which will serve as project justification in connection with the program and CON phases. Also, surveys by the JCAH and AOHA should be referenced for citations which could affect the need for a project. Examples of such types of violations and citations are as follows:

1. Dead end corridors in the surgery department violate life safety and fire codes
2. Lack of adequate space for separate handling of clean and contaminated supplies in the central processing department cited by JCAH or AOHA
3. Laboratory mechanical system which does not provide the amount of outside air required by the state fire marshall
4. Lack of direct observation of all beds in the intensive care unit was cited in the recent JCAH or AOHA survey

The need for a project because of poor existing functional configurations should have been established on a preliminary basis as part of the master facility plan (see Chapter II). During the program and project definition phase, the functional configuration would be studied more closely, state-of-the-art configurations would be evaluated, and the current and alternative operating systems within and between departments would be analyzed. The following are examples of need statements relative to functional configurations:

1. The medical beds are scattered and are not in an appropriate configuration or sufficient aggregate area for the development of an efficient nursing unit
2. The diagnostic rooms in radiology are arranged along a single corridor which does not permit an appropriate flow of patients and staff or efficient film processing
3. The movement of reprocessed, reusable items between surgery and central sterile supply is inefficient due to the distance

All statements of need and rationale for a project should be evaluated in the light of applicable guidelines and criteria of the appropriate government health planning and review agencies. As the hospital conducts its needs assessment, all documentation should be saved for use in the CON application and for reference during the design reviews by code officials.

TESTING DEMAND

Testing the projected demand for future services is one of the most important tasks at the program and project description stage, since service volumes determine space requirements and financial feasibility and affect CON approval. The following points should be considered relative to testing demand and the treatment of projected utilization statistics:

1. *Historical records*—The hospital's planning and finance departments should maintain current and uniform historical utilization statistics for inpatient nursing services:

   — licensed beds
   — average beds in use
   — average daily census
   — average length of stay
   — admissions or discharges
   — patient days

   and for major ancillary services (on both an inpatient and outpatient basis):

   — surgical procedures
   — radiology examinations
   — laboratory tests
   — deliveries
   — EKG and EEG tests

— physical therapy procedures
— occupational therapy procedures
— respiratory therapy procedures

2. *Planning base*—The same historical utilization statistics should be used for CON filings, space requirements planning, and financial feasibility projections. Statistics should be updated as necessary to accommodate the timing of work tasks.
3. *Assumptions for utilization projections*—Depending upon the age of the long-range plan, it may be necessary only to confirm the assumptions for the utilization projections. If, however, the most recently completed projections are not based on current data, all assumptions should be reviewed and revised as necessary. Important assumptions to consider include:

   — population growth
   — historical patient origin
   — service area competition
   — use rate
   — market share
   — service and program mix
   — payer mix
   — government planning agency guidelines and standards
4. *Bed need methodology*—Smaller construction projects may not require detailed bed need studies. For major projects, the approved bed need methodology promulgated by the appropriate government planning agencies should be referenced, as well as other valid methodologies developed within the health care industry in general.

Scope

The preliminary scope of the project is the physical description of the project in narrative, numeric, and/or visual form. (For CON purposes, a change in scope could be a change in the cost, space, or program. Scope has different significance with respect to construction contracts; a change in scope outside the contract could have cost implications.) For major projects, the scope definition should result in:

1. A listing of projected net and gross square footage by department based on:

   — existing space use
   — projected service volumes
   — current and future staffing patterns

2. A preliminary program narrative which describes department functions, departmental relationships, internal circulation patterns, and operating methods and systems
3. Conceptual drawings which depict the building additions and modifications in relationship to the site and departmental sizes and relationships

The above level of detail may not be necessary for smaller projects, which might only require the development of sketches and square footage projections for specific departments or areas. For areas to be remodeled, it is helpful to distinguish between heavy (gut and remodel), medium, and light (patch and paint) remodeling. For major construction projects involving several departments and significant remodeling, it is also helpful to document in tabular form the existing and future square footage for each department by building level.

BUDGET

Fairly detailed line item budgeting of project costs should occur as soon as the preliminary scope is defined. Major budget categories are as follows:

— New construction, renovation, and fixed equipment
— Movable equipment and furnishings
— Professional fees
— Administrative expenses
— Contingencies
— Financing costs
— Land and site improvements

The construction cost includes sitework, utility placements and landscaping, as well as the bricks, mortar, and engineering systems of the building. Fixed equipment consists of built-in or attached items furnished by the contractor, including cabinetry, casework, millwork, brackets, headwalls, surgical lights, and medical gas systems. For major construction projects, a hospital construction manager or an experienced cost estimator should provide the estimate for construction cost and fixed equipment. The level of measurable drawings and quantifiable information regarding the construction is very broad at the program stage. Thus, the program construction cost estimate should also be based upon a narrative description of significant building features and quantities, i.e., foundation system (caissons or spread footings), structural frame (precast concrete or brick), interior finishes (carpet or tile, paint or vinyl wall coverings). All of this information will assist the cost estimator in preparing a valid and reasonable budget.

Movable equipment and furnishings include radiographic equipment, movable medical equipment (monitors, crashcarts, film processors), furniture, and furnishings (patient beds, office desks, examination tables, graphics, and signage).

Individual fee budgets should be established for each professional discipline on the project team. Design fees for equipment, interiors, and change orders should be included. Administrative expenses is a suggested heading for a miscellaneous category to include soil tests, site survey, builders risk insurance, travel expenses, plan printing, building permit fees, etc.

Contingencies should include:

1. A cost estimating contingency, which may be included in the construction cost estimate
2. A design contingency, which allows for changes during design due to uncertainties at the program stage
3. A construction or change order contingency, which should be reserved until contracts are awarded, and which covers change orders during construction for errors and omissions on the plans, unknown field conditions, final code comments, and necessary but minor scope changes

For purposes of the construction contract, contingencies may have specific definitions. For example, a guaranteed maximum price contract may include specific references to bidding contingency, contracting contingency, and owners' contingency. (See Chapter IV.)

Financing costs vary depending on the type of transaction, but could include reserve funds, capitalized interest, bond discount or loan placement fee, and miscellaneous financing expenses (legal fees, authority and trustee fees, feasibility studies or letters, bond printing, title insurance). Additional detail regarding the budget for the financing plan is addressed in the following subsection.

Financing Plan and Feasibility

The financing plan for a debt transaction should be based upon the subtotal of all project costs *excluding* financing expenses, which are then determined as a function of the financing assumptions. Assumptions for the plan of financing include:

— Financing vehicle

— Schedules for design, financing, and construction

— Interest rate

— Reinvestment rate, if applicable
— Maturity of debt
— Percentage for placement fee or bond discount
— Capitalized interest period
— Cash draw down during construction
— Equity contribution

Based upon the plan of financing and estimated sources and uses of capital, financial statements may be projected to determine the financial feasibility of the proposed project. The following points should be considered in assessing whether the project is financially feasible and fits within the hospital's debt capacity:

— Projected operating income and excess revenues
— Projected debt service coverage
— Projected debt to equity ratio
— Projected change in working capital and cash
— Projected increase in rates and charges
— Compliance with the institution's internal financial guidelines

If the hospital recently completed a strategic financial plan or debt capacity analysis, it may be necessary only to test the program cost estimate and financing assumptions against the recent study. However, projected financial statements are generally required for CON applications.

Schedule

A detailed master project schedule integrating the work tasks of all professional disciplines for each project phase should be developed at the outset of the project at a team meeting. This schedule should be closely monitored and revised during the program and project definition phase, since it is a key assumption in the construction cost estimate and financing plan. Two key budget items which are a function of the schedule are construction escalation and capitalized interest.

Benchmark tasks in the master project schedule for the program and project definition stage include (traditionally in this order with some overlap):

— Complete utilization projections
— Complete preliminary program narrative
— Estimate future square footage

— Develop block drawings
— Estimate construction cost
— Prepare project budget excluding financing expenses
— Prepare plan of financing
— Complete projected financial statements
— Test project against hospital's debt capacity

Determination that the project is within the hospital's debt capacity is the last step in the project definition stage. Subsequent milestone events in the master project schedule are:

— Approval of project in concept by governing board
— Submit CON application
— Receive CON approval
— Approval by governing board to award architect/engineering contract
— Complete schematic design
— Schematic design cost estimate
— Complete design development
— Design development cost estimate
— Initiate financing activities
— Submit design for agency reviews
— Complete construction documents
— Release plans for bidding
— Agency reviews
— Receive construction permit
— Receipt of bids
— Complete all financing documentation
— Approval of financing by governing board
— Financing commitment (purchase offer or bond sale)
— Receive debt proceeds (closing of financing)
— Approval by governing board
— Award construction contracts.

VI

Certificate of Need Process

The information contained in this section is subject to change due to ongoing changes in the regulatory environment. As indicated previously, the hospital should reference the guidelines and regulations governing the CON process in connection with the project identification activities. CON research and analysis tasks should be incorporated in the master project schedule concurrent with the other activities of the program phase.

As each task in the program phase is completed, the associated data or documentation will be needed for a section of the CON application. Thus, the schedule may be dovetailed such that a draft CON application is completed soon after the project scope and budget are tested against the hospital's debt capacity.

Governing Board Approvals

Prior to filing a CON letter of intent or application, approval in concept should be obtained from the governing board, even though the subject capital expenditure may not occur for one to two years.

Federal and State Requirements

Federal

Public Law 92-603 was enacted in 1972 as part of the amendments to the Social Security Act. Included in P.L. 92-603 was establishment of the Section 1122 program. Under Section 1122, hospitals which incurred individual capital expenditures in excess of $100,000 were required to secure 1122 approval in order to be reimbursed for interest and depreciation expenses allocable to medicare patients. In 1983, Section 1122 was amended by P.L. 98-21 to change the threshold to $600,000 or such lower amount as the state may establish.

Public Law 93-641 was enacted in 1974 to establish the Certificate of Need program which provided for State Health Planning and Development Agencies (SHPDAs) and Health Systems Agencies (HSAs) to review and recommend to the Department of Health, Education, and Welfare (DHEW), now the Department of Health and Human Services (DHHS), concerning medicare reimbursement for capital expenditures. P.L. 93-641 required each SHPDA to administer a CON program which, unlike Section 1122, extended coverage to certain service changes whether or not a capital expenditure was involved. The law also required states to enforce the CON program; however, imposition of sanctions on the states for failure to conform to federal requirements has been deferred by Congress.

The Omnibus Budget Reconciliation Act of 1981 reduced the minimum annual federal grants to HSAs and also increased the federally accepted capital expenditure for CON review. The new federal threshold is $600,000 for new capital expenditures, $400,000 for major medical equipment, and $250,000 in annual operating costs for new institutional services.

State

Most states have CON legislation and continue to participate in the Section 1122 program. The capital expenditure thresholds established by the states vary significantly and in many cases are below the federal thresholds. Examples of different state limits are as follows:

In Michigan, the current threshold for capital expenditures and equipment is $150,000. All changes in institutional health services are subject to review.

Indiana's Determination of Need legislation sets the expenditure thresholds at $600,000 for capital expenditures, $400,000 for equipment, and $150,000 for annual operating costs.

Iowa's CON law establishes the same expenditure thresholds as those set by the Omnibus Budget Reconciliation Act. The Section 1122 limit in Iowa is $100,000, but review of projects under Section 1122 is usually waived for projects involving expenditures below the CON thresholds.

LETTER OF INTENT

From state to state, there are different filing requirements and time frames for submitting a letter or notice of intent prior to the time that a CON application may be submitted. Submitting the letter of intent should be included in the master schedule based upon the projected date for submitting the application and the advance filing requirements of the letter. Although the letter of intent is not binding, the details of the submittal should be kept as general as possible to allow the hospital some flexibility in developing the final scope and cost.

CON APPLICATION

One party should assume responsibility for scheduling and coordinating the input of the project team during the development of the CON application. This responsibility should include:

1. Distributing copies of the application form to all team members
2. Developing a task list of assignments, individual or firm responsibilities, and completion dates for all sections of the CON application
3. Assembling all information into the CON application form, reviewing the material for technical consistency, and adding to and editing narrative to achieve consistency of style and format
4. Advising the project team members on guidelines applicable to their respective areas of the application
5. Scheduling and conducting application review sessions by the project team
6. Communicating with CON review agencies
7. Submitting the notice of intent and final application

As indicated in the previous discussion regarding the program and project description phase, the work tasks pertaining to that phase should be scheduled to overlap with the schedule of tasks for developing the CON application. Thus, as projections of square footage and utilization are completed, these may be cast in the CON application format and compared with such CON guidelines as square footage per procedure, use rate, and projected bed need.

The greatest efficiency can be achieved in developing the CON application by concurrent scheduling of work tasks with the program phase and by interdisciplinary review of the draft application at milestone dates. For major projects, it is important that the team periodically review drafts of the application during its development so that areas of concern or major problems may be addressed well in advance of the scheduled submittal date.

Fiscal year rather than calendar year financial and utilization data should be used in the CON application so that the historical financial statements tie to the historical and projected financial data. This is important since the CON review agencies may calculate and compare such ratios as historical and projected cost per patient day.

It is important that the CON application include a thoughtful discussion of the manner in which the proposed project is consistent with the applicable government planning standards and guidelines. Federal, state, and local standards as indicated in the state health plan, approved bed need document, health systems plan, etc. should be addressed. Areas where the project is inconsistent with the guidelines should be explained. Assistance from legal counsel regarding the status of regulations and interpretation of guidelines documents may be required. For controversial projects, review of the CON application by legal counsel prior to filing may be desirable.

Prior to submitting the application, it may be helpful to review the final draft with staff of the applicable review agencies for comments and to identify potential problem areas.

HEARINGS

Once the application is submitted, there is a period for completeness review by the staff of the state and local health agencies. Once the application is deemed complete, the review hearings, if any, are scheduled. The party responsible for developing the CON application should also organize the project team for the CON hearings. This would include:

1. Circulating membership lists of the CON review board to the project team and identifying friendly and hostile votes
2. Circulating hearing schedules to the project team
3. Determining project team members' attendance at the CON hearings
4. Circulating copies of any staff reports prepared by the review agencies to the project team and coordinating responses from the team to areas of concern in the staff analysis
5. Organizing presentations, if any, to review boards, including selection of speakers and preparation of presentation outline and audio-visual materials
6. Preparing hypothetical questions and answers which may arise during the hearing

7. Organizing the presentation rehearsal and mock question and answer session with the speakers and other team members

Generally, hospital administration rather than consultants should make presentations for CON hearings. Administration should be prepared to answer all major questions to the extent possible, but should not refrain from directing a technical question to a consultant in the audience if the point is important. The decision whether to have consultants at the hearing depends on the size and nature of the project. Certainly all resources should be available for major projects. For controversial projects, legal counsel should attend the CON hearings to monitor the proceedings for illegal procedures which could be used in a CON appeal.

It will be important to demonstrate community support for the project. Letters of support for the project should be solicited well in advance of the CON hearings. The hospital should take care in recruiting and preparing community support and testimony for the CON hearings so that it does not appear to be orchestrated.

VII

Design Phase

This chapter addresses detailed design issues, the role of hospital staff during design, the various construction cost estimates at each design phase, monitoring the overall project budget and master project schedule during design, and fast tracking design.

General Approach

Chapter III has a discussion of the following five basic design phases: (1) program narrative and block drawings, (2) schematic design, (3) design development, (4) construction documents, and (5) record drawings. Generally, the program narrative and block drawings phase will have been completed for the CON application. Then, the design team activities may be put on hold until CON approval is received, since initiating schematic design could entail significant fees.

In certain situations, it may be determined to initiate schematic design prior to the receipt of the final CON approval, i.e., for small projects, upon receipt of a favorable CON staff analysis, subsequent to a favorable preliminary CON hearing, and/or in times of high construction inflation. This overlapping allows compression of the overall project schedule which could result in accelerating the date for completion of construction.

This approach, however, involves risks that the CON approval will not be received and thus the incurred design fees above CON approval threshholds may not be reimbursed by third party payers. Another risk in overlapping the design and CON approval phases is that final settlement of CON approval may require modifications of the original proposal. In this case, part or all of the complete schematic design based on the original project would be wasted, possibly resulting in additional design fees.

For certain construction projects, particularly for major programs or those involving several departments, the preliminary program narrative must be expanded to a detailed functional program prior to initiating schematic design. This task could require 30 to 90 days to complete, but schematic design may be initiated sooner to overlap somewhat with the programming. As design proceeds, the functional program will be modified and refined to reflect changes in the plans.

The hospital should recognize that there will be changes during design as the plans become more detailed and as hospital staff at the department level and physicians review their particular areas. The ease with which design proceeds and problems are encountered will depend largely on the quality of the preliminary planning underlying the project. As conditions and issues unknown or unsettled at the program stage become identified and resolved during the design stage, the hospital must be flexible in recognizing and adapting to this dynamic process. A challenging and refinement of earlier thinking is a necessary and healthy function of the design process.

THE ROLE OF HOSPITAL STAFF

It is important that the CEO not completely delegate the completion of design activities to other hospital staff, consultants, and the architect. Numerous points affecting larger issues of the institution could arise during the design process, and the CEO should be aware of these points. During this process, the monthly project team meeting should be the forum to report on the status of design to the CEO and other hospital staff who are not involved in the detailed design sessions.

The CEO or the project manager should designate a hospital administrator who will oversee the design process relative to hospital staff reviews. This member should attend design review sessions between the architect, the facilities planner, and department managers for all major affected areas. This administrator may have to arbitrate between other hospital staff when conflicts arise concerning design issues or should design sacrifices be required due to budget constraints. It is important that the architect not be forced to play this role; otherwise, the staff could feel that a particular design was forced upon them by an outsider rather than selected by the senior administration as the best alternative to meet overall project goals.

All departments affected by the project, whether directly or indirectly (e.g. ancillary and support departments), should review the preliminary design. During schematic design and design development, the architect should hold several design work sessions in which each departmental area is reviewed in detail with the individual department manager, nursing administrator, clinical director, and other involved management staff. Key physicians of different specialties should also be invited to these review sessions to provide medical staff input. The design work sessions are generally scheduled for a full week every six to eight weeks until construction documents are initiated. The work sessions involve intensive challenging of the plans and discussion of the design concepts to assure that the architect properly understands the staff's requirements for layouts, flows, storage, etc.

The administrative staff member who is responsible for supervising the design process should assure that the staff are aware of their review schedules, that all important staff are available or can provide input through others, and that all staff have the plans with adequate advance time for review. It must be stressed to department managers that their thorough review is critical to avoid errors and omissions later which could result in change orders. Departmental reviews include checking the number and location of outlets, door opening directions and widths, lighting plans, casework and millwork type, function and location, equipment layout, partition locations, etc. At the conclusion of each onsite design work session, the architect should meet with all senior administrative staff, including the CEO, and all key department managers to summarize that week's sessions, indicate pending design problems and define the schedule and level of review for the next work session. During design work sessions, morale in specific departments could drop should budget constraints force design sacrifices or should other design problems occur. Including key department managers in the concluding sessions helps to boost the sagging morale and regenerate enthusiasm as the various managers hear about the particular design problems of other departments, and the progress of the total project.

Once construction documents are initiated, the input of department managers should be essentially complete. The engineering systems become the critical design feature during the construction documents phase when the hospital engineer should carefully scrutinize the drawings for consistency, coordination among systems, and relevance to operating and maintenance issues. The coordination of the plans is important since each of the design and engineering disciplines are developed separately and then melded together. For example, the mechanical drawings are overlaid on the architectural backgrounds, then the electrical drawings are overlaid upon that, etc.

Design Phase Cost Estimates

The construction manager or the cost consultant should provide cost estimates for construction and fixed equipment at the completion of each design phase and check for cost increases during each phase. The program cost estimate which is used in the CON application was discussed in Chapter V. The schematic design and design development construction cost estimates should theoretically be more accurate than the first estimate since the plans are more detailed and at a larger scale. Thus, the hospital should not be surprised should these later estimates exceed the program budget, due to the more detailed plans and ensuing progress of design. In the case of a budget overage, the task of the design team is to identify alternative design solutions, materials, systems, and construction methods which could reduce the budget.

When a particular phase of design is complete and the plans are handed off to the construction manager for estimating, the hospital must decide whether to proceed immediately with the next design phase or delay it until the cost estimate for the prior phase is complete. For major projects, the estimating process may require four to six weeks. Given this time frame, the hospital may wish not to interrupt the design, but rather, to continue the design and risk backtracking in four to six weeks should the cost estimate for the prior phase require reductions and scope modifications. The decision to delay the next design phase for the cost estimate will depend upon factors such as construction escalation and the overall master project schedule.

Upon completion of the construction cost estimate for each design phase, the hospital staff, the architect, and the facilities planner should meet with the construction manager for a line item review of the estimate. The construction manager or cost consultant should be directed to provide detailed documentation regarding:

- Costs and square footage by department
- Costs and square footage by distinct building category, if more than one
- Costs for each trade category
- Unit prices for materials
- Quantities of materials
- Adjustments for regional wage scales and renegotiation of labor contracts
- Escalation assumptions
- Schedule assumptions
- Construction managers's construction phase fee
- General condition items

The first run of each estimate often contains errors, omissions, or duplications and should be carefully reviewed by the design team and revised prior to presentation to the project committee or governing board. During the design team's review, the estimate for each department and building category should be compared with that for the previous estimate to determine changes from the previous design phase.

Whether or not the cost estimate exceeds the approved budget, the construction manager should conduct value engineering exercises to suggest changes in systems, materials, and construction methods to reduce the cost. Value engineering obviously is most important when the estimate is over budget, but the hospital must be careful not to accept any suggested reduction until the architect, the engineer, the facilities planner, and the hospital staff can carefully evaluate the alternative.

Some types of cost reductions can be designed as add alternates; other changes are irrevocable. An example of this is reducing square footage. This is irrevocable since the foundation and structural system are designed as a function of the building footprint which is predicated on the total square footage. Add alternates are features which can be separately designed and bid and then added to the scope of the project should the final bids be below the established budget. Examples of potential add alternates include vinyl wall covering in lieu of paint, pneumatic tube system, upgraded boiler capacity, and medium rather than light remodeling of a particular area. Alternates should be developed and assigned priority during design, but the number of alternates should not be so extensive that significant additional design fees are incurred.

The original project budget should include a design contingency so that uncertainties or changes during design which result in cost increases can be accommodated in the budget without compromising the project scope. Even given this contingency, the design team should continue value engineering exercises to achieve the best building for the price. It is not necessary to spend the design contingency which could be added to the change order contingency or deleted entirely.

To accelerate the overall project schedule, the cost consultant's or construction manager's final cost estimate based on construction documents should be prepared concurrently with the bidding period. The contractors' prices may then be compared with the estimate to determine the fairness and competitiveness of the bids.

Monitoring the Budget

The design process may extend over a long period of time; often 18 months or even longer are required to design a complicated building project. During this period there are many changes that could occur and new information that could be identified which may affect the budget. For example, midway through design development, it may become apparent that the hospital must complete certain utility relocations which are essential to the project but which have been excluded from the construction budget. The budget for all project cost items should be closely monitored during design so that no surprises arise at the point of financing, award of construction contracts, or during the construction period.

A log of purchase orders and actual expenditures during the prefinancing and preconstruction periods should be maintained, and it should be in the format of detailed line items within each budget category. Preconstruction expenditures could include planning and design fees, soil tests and borings, surveys, blueprints, and land acquisition. Examples of preconstruction purchase orders could be equipment and furnishings with long lead procurement times or which may be planned for early purchase, or for installation in existing facilities and then future relocation to the new or remodeled facility.

It is particularly important to monitor project cost items for which the initial budget was established based upon a parameter estimate rather than a firm price. For example, while the facilities planner's fee may be a firm percentage of the construction cost, the budget for the associated reimbursable travel expenses may be just an estimate. Monthly monitoring of the budgeted to actual reimbursable costs is therefore essential to maintain control of the overall budget. This is also applicable to consultants and attorneys whose fees are based upon hourly or per diem rates, since again a fixed price is not established.

The architect and the facilities planner generally base their fee invoices on the amount of design completed to date. The invoices should be checked to make sure that the percentage fee billed is consistent with the percentage of completed design. Major projects involve significant design fees and, for example, a 5 percent acceleration of the design development fee on a $10 million project could result in an overbilling of $10,000 to $15,000. Careful review of invoices can result in billing modifications which will enhance the hospital's cash flow.

A helpful tool for monitoring the budget is projected cash flows of project cost expenditures. The cash flow should be on a monthly basis, and separate projections should be developed for each budget line item, all professionals within the fee budget, and all architectural services within the A/E fee category. The projected cash flow is also useful in developing the plan of financing since the projection indicates total expenditures prior to financing and subsequent monthly draws upon which to estimate future interest income, if applicable.

MONITORING THE SCHEDULE

Work activities which are a function of other work tasks are said to be on the critical path of the schedule. Items on the critical path should be identified so that delays during design may be evaluated for their impact on the construction schedule and budget.

Monitoring the schedule during design is a key concern since it affects the construction escalation assumption and capitalized interest assumption, both of which affect the budget. Slipping the design schedule could significantly affect the construction schedule due to seasonal conditions, i.e., the ground breaking may be planned in June to allow the new building to be enclosed by November so that work can continue through the winter. A two month design delay could result in a four to six month construction delay if outdoor work must stop due to inclement weather.

A key assumption in the financing plan is capitalized interest, which is explained in detail in Chapter IX. Since capitalized interest is a function of the construction duration, any change in the construction schedule must be communicated to the financing team members immediately. Then the financing plan may be revised and any effect on the total financing costs may be evaluated.

The master project schedule should be expanded prior to each new project phase to include all detailed work tasks for that phase. This schedule should be referenced periodically and the team members contacted to assure that task accomplishment is timely.

The status of the schedule should be an agenda item for all monthly project team meetings. In addition, there should be follow-up on the schedule between meetings. This is necessary since the work during certain phases can be detailed, fast, and furious, and it is easy for completion of a specific task to "fall through the cracks" or run behind schedule. Since many events dovetail in the master project schedule, there can be a domino effect on other work tasks should one task slip. For example:

1. The soils report is necessary to complete the foundation scheme, but the report is two weeks late.
2. The foundation scheme is necessary to complete the design development package (due February 1st) for the construction cost estimate. Lacking the soils report, the structural engineer was assigned to another project for two weeks and then lost three more days to complete that task and get reoriented to the hospital's project.
3. The construction manager received the design development package two weeks and three days late. Due to other demands for the CM's cost estimating staff and computer time, another four days was lost in preparing the design development (DD) cost estimate.

4. The DD cost estimate was originally scheduled for presentation to the project committee March 15th but is not completed until April 6th. The estimate is over budget and the project committee must make a decision regarding scope reductions on April 15th. The project team has only a week to prepare and distribute a comprehensive report to the committee.

   Meanwhile, construction documents have been in process for two months and the necessary scope reductions could affect this work.

The keys to effective schedule control are projecting in advance the detailed work tasks, confirming in writing all schedule commitments, and interacting and communicating with the project team members between monthly meetings. Concurrent rather than sequential scheduling of work tasks will compress the overall project schedule but requires close monitoring and coordination of the individual team disciplines.

FAST TRACKING DESIGN

Because fast tracking or phasing design may be desirable in order to complete an early construction phase, a general discussion was included in Chapter IV in relation to alternative construction approaches. Under the fast track design approach, the plans for such early work as the excavation, foundation and/or structural system may be accelerated to be complete prior to the remainder of design.

Fast tracking design requires close review of the structural engineering plans and coordination with the incomplete architectural backgrounds and layouts. Items not yet on the plans must be visualized to assure that the completed structural and other drawings account for the potential effect of the missing items. Fast tracking design involves the risk that additional change orders to the early construction phase will be required as a result of the final plans for the entire package.

It may be difficult to secure building or construction permits for the early phase since many review agencies require substantially complete plans. This should be confirmed prior to finalizing the fast track schedule. It is not prudent to fast track design unless an early construction phase can be commenced, since completion of partial plans will be of no value and will only complicate the engineering coordination task.

EQUIPMENT PLANNING

Planning for fixed and movable medical and nonmedical equipment, radiographic equipment, other specialty equipment, and furniture and furnishings should be integrated with the building design activities.

Depending on the size and complexity of the project, it may be advisable to engage an equipment planner and/or interior designer. Many architectural and hospital facilities planning firms also provide services in medical equipment planning.

Both the architect and the facilities planner will advise on the appropriate schedule for integrating the equipment planning work into the various phases of design. Detailed equipment planning should be initiated no later than the beginning of the construction documents phase since the design specifications, i.e., ceiling heights, room layouts, electrical, and gas hook ups, could be affected. This is particularly true for major medical equipment. For budgeting purposes, equipment planning based upon general parameters should occur during the early design phases.

The procurement and installation of equipment must be carefully planned and scheduled. Either an equipment consultant or a hospital staff member should have responsibility for the equipment planning, procurement, and installation process.

VIII

Agency Reviews and Bidding

The agency review period consists of securing design approvals from the various state and local agencies and the fire marshall in order to obtain a building permit. The agency reviews and bidding phases may overlap each other, and the schedule of the agency reviews may be partially concurrent with the final design phase, as indicated in Chapter III. The following subsections address in greater detail the strategies, timing, and other criteria important to the review and bidding process.

Local and State Code and Fire Agencies

At the start of the design process, the hospital staff and the architect should develop the schedules and a list of all agencies who must review the design. During each design phase, meetings should be held with the staff of these agencies to solicit their comments on the plans to date. A reviewing agency may also require the hospital to submit a functional narrative or operations manual which describes how the building will work.

Interpretations of the building codes may vary from agency to agency and even from individual to individual. For use in review sessions and future negotiations, the hospital should request agency staff comments in writing. The hospital should also seek waivers or appeal code interpretations in areas of disagreement which have major schedule or cost implications.

It is important to recognize that the fire marshall is independent from the agency of the state health department which reviews design. The hospital may be less successful in negotiating with and appealing to higher authorities in connection with the fire safety plan review. The fire marshall's approval is generally required before the applicable state or local authority will release a construction permit.

It is almost inevitable that major construction projects will entail final design revisions as a result of the agency review process.

TIMING

Chapter III identified three approaches to timing the agency review and bidding process. In determining whether to overlap the final design, agency review period, and bidding phase, the hospital should assess the complexity of the plans, the posture of the review agencies, the bidding climate, and the master project schedule. Should it be determined that concurrent scheduling is advisable to accelerate the total schedule, great care must be taken to coordinate the activities, communicate with the contractors, and if applicable, communicate with the financing team.

Should the hospital decide to overlap the agency reviews with the bidding process, contractors should be advised that midway through the bid preparation period design addenda may be forthcoming.

ADDENDA

Design addenda are supplements to the completed design package which contain additions or modifications. Addenda may be distributed to contractors during the bid preparation period as a result of final comments from the review agencies or corrections arising from the final design quality control and check out by the design team.

It is important that addenda be appropriately handled with the contractors so as not to confuse or pressure them. Adequate time should be allowed before the bids are due so that the contractors can carefully evaluate the price of the addenda. A significant number of addenda or unclear or complicated additions can frustrate the bidders who may react by padding their bid to cover uncertainties, delaying their bid submittal, or not bidding at all. Since the object of the competitive bidding process is to obtain the lowest price possible, it is best to maintain a good relationship with the contractors during the bidding period. The hospital should exercise discretion in deciding to process design revisions either as addenda to the bid documents or as change orders to the final awarded construction contracts.

The effect of design changes tend to ripple through the plans, since one change in the location of a mechanical duct could change the piping lengths, electrical conduit location, outlet size, door opening, etc. Therefore, the addenda must be carefully coordinated to assure that all affected areas of the plans are covered.

FRONT ENDS

The plans and specifications upon which the bidding contractors calculate their prices include drawings and narrative descriptions as well. This narrative material could be called "front ends." It includes the general and special conditions of the contract, the specific phasing schedule for construction, the payment procedure and schedule to be followed, requirements for submitting waivers of liens, required construction methods, safety and insurance requirements, and the form of the contract with the subcontractors. Other narrative material which should accompany the bid documents includes outline specifications and finish schedules which consist of hardware descriptions; quality of such interior finishes as carpet, paint, wall covering, and tile; glass types; and specifications for performance of engineering systems.

In addition, the narrative information in the bid documents should include instructions to bidders regarding the format and schedule for bidding, bid bond requirements, conditions precedent to awarding contracts, and other data which could affect the final bids of the contractors.

Front-end information should be assembled during the design phase for close review by the entire project team. Any identified constraint which could affect the construction process should be included in the front-end documents. For example, should it be necessary to maintain the operation of the radiology department during its remodeling, contractors must be prepared to include in their bids premium costs for labor overtime to complete the renovation during hours other than the usual work day.

Legal counsel and the finance staff should review the portion of the narrative front-end documents which pertain to payment procedures, bid bonds, etc., to assure that the specifications meet the needs of the financing, if applicable.

PREQUALIFYING THE BIDDERS

The hospital may elect to issue an open invitation for bids, or may invite only qualified bidders to submit bids. Prior to distributing the bid documents to interested contractors, the construction manager, the architect, and the hospital staff could develop a list of candidates for bidding and prequalify the candidates according to the following criteria:

1. Use of union or nonunion labor or both depending on the degree of unionization of hospital personnel and the local climate relative to unions
2. Experience with specific contractors on previous projects
3. Ability of the contractors to meet the schedule or conform to required construction methods
4. Ability of contractor to secure a performance bond
5. Preferences for local or state firms

In order to obtain the most competitive prices, it is best not to exclude nonlocal firms or nonunion labor.

As an alternative to prequalifying the bidders, the bid requirements may be published for public bidding, and then the hospital must check each received bid against the requirements.

Prebid Conference

For major or complex construction projects, a prebid conference with the contractors may be scheduled on the site two to three weeks prior to the date that bids are due. The architect, the engineer, the construction manager, and appropriate hospital staff are present to review the bidding format and requirements with the contractors and to answer any questions regarding the plans and specifications. Also, at this time, addenda may be distributed or further explained. A small turnout at the prebid conference may indicate lean contractor interest which could suggest additional efforts to stimulate bidder interest. Should the contractors express significant concern or misunderstanding regarding the plans at the prebid conference, the hospital may consider extending the date that bids are due to encourage more accurate prices.

Receipt and Analysis of Bids

The instructions to bidders should include the date, time, location, and contact person for the submittal of bids. Telephone and late bids should be discouraged to ensure fairness to the conforming contractors, but the hospital may wish to further investigate the telephone or late bids if the first bids are disappointing.

Bids should be opened privately by the hospital family and/or project team when possible, since the construction manager may wish to subsequently negotiate with the low bidder relative to better prices for alternates. In such a case, the hospital would have more leverage if the contractor were not aware of the competition's prices. In a construction management situation, the contract with the CM may govern the owner's rights relative to the bid opening.

Upon receipt of all bids, the prices should be tabulated by category and any clarifications to the bid should be noted. Under a general contractor arrangement, the single low bid would be identified. For multiple subcontractors, the low bids for each category are summed to arrive at the base construction cost.

To preserve the integrity of the bidding process, it is advisable to accept the low bid of the qualified bidders. Particular circumstances may influence the decision not to accept the low bid, i.e., local versus nonlocal contractors. In such a case the hospital could be subject to criticism in the contracting community and this could affect the results of later bid phases. It is recommended that, once bidders are prequalified and bid specifications established, the low bid should be awarded the contract on the basis of fair competiton.

BID BOND

The contractor may be required to submit a bond with his bid which will guarantee his acceptance of the contract at the price submitted. The bid bond usually can be returnable if the contractor is unsuccessful.

PERFORMANCE BOND

Generally it is required that the contractor submit a bond which guarantees his ability to complete the work for the contract price. Should the contractor be unable to complete the work, the hospital would be protected by the bond for the outstanding cost of the work left incomplete. The size of the performance bond should be determined by the hospital. The requirements for the performance bond should be stated in the instructions to bidders and the bid analysis should include checking the bond.

ALTERNATES

Alternates are design features which can be separately designed and bid and then added to or deleted from the scope as the budget dictates. Generally, add alternates are preferred to deduct alternates because the contractors may provide better prices when their base profit is calculable.

All contractors may choose not to bid on all alternates. Some contractors may not be interested and some alternates may apply only to a limited number of trade categories. Certain bids on alternates may stipulate a deadline for accepting the alternate. Upon analysis of the bids to insure that all low bidders are qualified to complete the work, should the bids be under budget, the alternates may be reviewed and priorities confirmed or modified given the actual alternate prices. The hospital and project team may then recommend to the project committee that certain alternates be accepted at the same time that the contract awards are made.

IX

Financing

The financing process is a critical element to the successful implementation of a construction project. Whether or not a long-term debt transaction is involved, the strategy for funding the capital expenditure should be well defined and integrated with the overall project development process. This chapter includes discussions of alternative financing vehicles, preparing the plan of financing, financial guidelines, finance authorities, schedule, legal documents, feasibility study, and governmental and governance approvals.

Alternative Financing Vehicles

Chapter III introduced the following types of financing vehicles—tax-exempt loans, taxable loans, FHA-242, and interim loans.

The volatility of the capital markets in recent years has stimulated the development of esoteric financing mechanisms not traditionally applied to the hospital industry. Such alternatives include variable rate demand notes, lower floater bonds, and quarterly tender bonds. These financing alternatives avoid long-term commitments at high interest rates, but have inherent risks regarding future availability of capital at rates below the current market. Other financing alternatives, including off balance sheet financing and developer financing, may be desirable debt vehicles to preserve the hospital's debt capacity and may be appropriate for certain types of projects, including parking garages and medical office buildings. A joint venture limited partnership with physicians may be an off balance sheet financing mechanism to utilize outside capital to preserve hospital debt capacity. Additional references should be consulted for a detailed review of these financing vehicles. The traditional hospital financing techniques and methods are described in the following sections.

Tax-Exempt Bonds

The publicly issued tax-exempt bond is a popular financing mechanism used frequently by the hospital industry for major capital construction projects. The bonds are generally secured by a lease or mortgage on the financed assets and a pledge of the hospital's revenues to pay the debt service on the bonds. Tax-exempt revenue bonds may be sold on the open market to the public or placed privately with a bond fund, bank, insurance company, or small group of investors.

Tax-exempt bonds must be issued through a municipality or quasi-public agency at the municipal or state level, depending on applicable local law. This is commonly known as the issuing authority. Some states do not have state authorities but have local authorities. In certain states, municipalities can act as issuing authorities.

Another tax-exempt loan vehicle is an industrial development revenue bond which may be issued in situations where existing legislation does not provide specifically for the issuance of tax-exempt bonds for hospitals. Legal counsel should be consulted regarding the availability of industrial development revenue bonds when a local hospital finance authority is not an option.

Tax-exempt financing is a desirable financing mechanism since the interest rates are below taxable rates. The spread between long-term tax-exempt and taxable rates has ranged from 1.5 percent to 3 percent in recent years. The business covenants in the legal documents which govern tax-exempt financings are also generally more flexible than for taxable transactions. The legal documentation is further addressed in a later subsection.

Bond insurance may be available depending on the institution's credit rating. An insured bond receives the highest possible credit rating, which reduces the interest rate on the debt. However, bond insurance is not necessarily the most economic option. Its value depends on the present value of the insurance premium cost as compared with the present value of future reduced interest payments.

Taxable Loans

Taxable loans include:

- Direct loan with a bank or insurance company
- Private placement with a group of investors, banks, or bond funds
- Public bond or note issue

Taxable loans are generally secured by a mortgage on the property of the financed hospital. Taxable loans may be simpler to structure than tax-exempt financings which require significant preparation and documentation. To the borrower, a taxable loan means higher interest rates and generally more restrictive legal covenants than for a tax-exempt loan. Often, taxable lenders prefer to limit the loan term to 20 years. A balloon payment or a large principal payment at maturity is a common feature of certain types of taxable financings.

FHA-242 Loan

A Federal Housing Administration FHA-242 loan is structured with a federal interest subsidy in connection with a taxable loan from a mortgage bank. FHA-242 loans are secured by a first mortgage of the financed hospital's property. An FHA loan may not be the preferred financing vehicle for those institutions with strong credit who have good access to the capital markets for the following reasons: the application procedure for such a loan is lengthy, the design of the project must meet certain federal standards for construction, and the legal covenants associated with FHA loans are often onerous.

A loan guarantee by the Government National Mortgage Association may be used in combination with an FHA-242 transaction to bolster a weak credit.

Interim Loans

Interim financing, as distinguished from long-term financing, is a source of funds to cover costs for a short period of time until the permanent financing transaction can be completed. If it is necessary to defer the permanent financing until firm bids are received or until the long-term capital market improves, an interim loan may be required to cover project expenditures for fees and early construction phases.

In certain situations, it may be desirable to complete the entire construction project with an interim loan, which would then be permanently financed on a long-term basis upon the completion of construction. Generally, it is assumed that interim loans for capital projects will be refinanced on a long-term basis since the long debt service payment schedule matches more closely the depreciable life of the financed asset than does a short-term repayment.

An interim loan could be in the form of a line of credit whereby a total dollar commitment is made and fees are paid based upon the amount of the commitment; however, only the actual level of funds required each month are drawn down from the lender, and the interest costs actually incurred are based upon this total principal draw.

PREPARING THE PLAN OF FINANCING

Chapter V indicated that the plan of financing could include the following assumptions:

- — Financing vehicle
- — Schedules for design, financing, and construction
- — Interest rate
- — Reinvestment rate
- — Maturity of debt
- — Principal amortization
- — Placement fee or bond discount
- — Capitalized interest
- — Cash draw down during construction
- — Equity contribution
- — Miscellaneous financing expenses
- — Debt service reserve fund

Many of the above factors will not be relevant to a capital expenditure which involves no debt. The following descriptions of the above financing assumptions are germane to larger projects involving debt financings.

Financing Vehicle. Alternative financing vehicles were discussed in the previous section. It is important to define the particular vehicle prior to attempting any calculations of financing costs or the financing schedule, since each vehicle has different characteristics which affect the other financing assumptions.

Schedules. An integrated schedule for design, financing, and construction must be established prior to finalizing the remaining financing assumptions, since many assumptions are a function of the schedule, including capitalized interest and principal amortization. For the plan of financing, milestone events in the project schedule include:

- — Closing of financing
- — Initiation of construction
- — Date of the bonds or notes
- — Completion of construction for each distinct building phase
- — Dates to initiate funding for principal and interest payments
- — Dates of first principal and interest payments
- — Date of final debt service payment

Interest Rate. Serially structured loans may have multiple interest rates associated with the different maturities. In order to calculate the financing plan, an average annual interest cost should be estimated. Since interest rates can fluctuate significantly during a given time period, the current rates should be checked periodically during the design period and the proposed financing plan revised accordingly.

Reinvestment Rate. In certain transactions, particularly for tax-exempt bonds, certain funds are established to be held and reinvested by the bond trustee until needed. These funds are:

— Debt service reserve fund

— Construction fund

— Capitalized interest account

The debt service reserve is a requirement of most public tax-exempt bond issues and is typically equal to the maximum annual debt service on the bonds. This reserve is held by the bond trustee for the term of the loan and is used to pay debt service in the event of default. The debt service reserve fund is also used to make the final debt service payment at the maturity of the debt. The construction fund contains money for the construction, equipment, fees, contingencies, and miscellaneous project costs. The capitalized interest account consists of funds to cover interest payments on the bonds during construction.

Until these funds are needed, they are invested in qualified securities and earn interest which usually flows to the construction fund. The reinvestment rate is thus the average rate that these funds will yield. The interest income during construction is calculated in the plan of financing as an available source of funds to pay project costs.

Maturity. The maturity of the debt is the term that the loan is outstanding from the original issue date. It is important to specify the term since it governs the amount of annual debt service.

Principal Amortization. It is traditional in hospital finance to defer principal payments on a loan until the capital construction is completed and the cash flow from third party reimbursement for depreciation expense associated with the expenditure is initiated. Thus the principal amortization period is the loan maturity less the construction period. Should the start of construction be different than the date of financing, however, the period of time from the date of financing to the completion of construction would be the period during which no principal is amortized. Generally principal payments are made on the anniversary date of the closing of the loan.

Placement Fee or Bond Discount. Compensation to the lender bank or underwriter for placing the loan or managing and selling the bonds must be included as a use of funds in the plan of financing. Investment bankers' compensation for managing and selling bond issues is expressed either as a percentage of the total issue or fee per thousand dollar bond. Fees for private placements and direct taxable loans are generally less than for public tax-exempt issues.

Capitalized Interest. Third party payers do not reimburse for interest expense on loans until the facility associated with the loan proceeds is operating. For this reason, the interest to be paid during construction must be capitalized. For tax-exempt bond issues, that amount of interest is generally borrowed as part of the bond issue. For loans which are drawn down as needed to pay construction, the accrued capitalized interest associated with each principal draw is added to the final principal balance of the loan.

In calculating the plan of financing, depending on the nature of the project, it could be advisable to add a cushion to the capitalized interest period. This would protect the hospital against a construction delay due to strike or inclement weather, or it might be applicable to a new facility startup situation.

Cash Draw Down During Construction. An estimate of monthly payments of project costs during construction is necessary to calculate the interest income on the construction fund. The construction manager or the contractor should calculate the construction cost draw down schedule, to which must be added the expected payment schedule for equipment, fees, etc. The projection of the total cash draw down should be coordinated by the project manager. This projection is also necessary for the trustee at the time that the debt proceeds are invested, to assure that the maturities of the securities match the project cost draw down schedule.

Equity Contribution. Generally all project equity must be paid prior to the closing of financing. Should it be planned that equity from operations or from payments on pledges be contributed during the construction period, the hospital may be required at the closing of financing to secure the future payment of equity with a letter of credit. In calculating the plan of financing, it is important to address the timing of the equity contribution. All costs paid for fees, etc., prior to the closing of financing may be counted as project equity. In some cases prepaid costs could exceed the proposed equity amount resulting in reimbursement to the hospital for the overpayment from bond proceeds.

Miscellaneous Financing Expenses. The plan of financing should include miscellaneous financing fees and expenses to the extent not covered in other fee budgets. Financing expenses include costs for bond printing, official statement printing, bond counsel, hospital counsel, feasibility studies and reports, trustee fee, issuing authority fees and expenses, title insurance, financial advisor, and rating agencies.

SCHEDULE

The financing schedule must be tailored to meet the needs of the construction schedule and, to a certain extent, vice versa. As explained in Chapter V, the schedule developed at the CON and project definition phase must include an assessment of the timing of financing if a debt transaction is involved, since this leads to the capitalized interest calculation which must be included in total project costs for the CON budget.

Generally, the permanent financing is not completed until either firm bids or an early GMP are in hand (see Chapter III). Particularly for tax-exempt transactions, legal preferences are that the construction cost be fixed prior to selling the bonds to assure that adequate funds to complete the project have been provided. Hence, the financing should occur after the receipt of bids. However certain intervening tasks must be scheduled between the receipt of bids and the closing of financing. These tasks include finalizing the feasibility study, marketing the debt issue, processing the bond rating, securing final approval from the issuing authority, and finalizing all legal documents, certifications, and required filings.

The construction manager should be aware that the commencement of construction may not be possible immediately after bids are received. A one- to two-month period may be required in the schedule to complete financing activities.

Early input from the finance team regarding the timing of the financing is important. It is helpful to keep the finance team apprised of the overall project schedule during the design period so that as much up front work as possible can be accomplished on the financing in preparation for the date that funds are needed to start construction.

LEGAL DOCUMENTS

Every debt transaction involves legal documents since the hospital must enter into a contract with the lender for the use and repayment of the borrowed funds. As indicated earlier, alternative financing vehicles entail different types of legal documents and covenants.

The legal financing documents contain many business covenants which affect the operations of the financed hospital. Important covenants which should be noted include:

1. Rate covenant, which sets a minimum required debt service coverage and thus indirectly the level of rates which must be charged
2. Permitted indebtedness which limits types and levels of additional debt
3. Insurance requirements
4. Merger or acquisition restrictions
5. Lease and sublease provisions
6. Disposal of assets

FEASIBILITY STUDY

Generally a feasibility study is required to market a public tax-exempt bond issue. The purpose of the feasibility study is to provide potential bond buyers with an independent assessment of the hospital's future ability to cover debt service. The feasibility consultant will use conservative assumptions in developing the study with respect to both demand and financial operations. The hospital should be aggressive in challenging the assumptions and results of the study during its drafting stage but should not expect the study to reflect unsubstantiated projections of significant volume growth. The feasibility study should be scheduled for completion at the time firm construction bids or a guaranteed maximum price are available. The consultant generally will not release the signed study until construction costs are firm because this is the single largest assumption in the study. The study should not be completed too far in advance of the debt marketing since it may grow "cold" and require updating prior to release.

X

Construction

Once all financing commitments are received or funds are actually in hand, construction contracts may be awarded. Generally a financing commitment should precede the contract award to assure that all funds necessary to complete the contract are available. Alternatively, the contracts may be awarded contingent on completion of the financing. The following subsections address specific issues which are important to the construction phase.

Contract Award

Upon receipt of bids and governing board approvals, the contract can be awarded immediately. The actual preparation and processing of the construction contracts could take several weeks to complete. This is particularly true in a construction manager agency relationship where multiple subcontracts are involved. The time frame is dependent on the location of a contractor's home office and number of required reviews at that level. Legal counsel should always review the language of the contracts prior to bidding. Some early construction activities may be commenced while the contracts are being finalized. In such a case, a letter of intent to sign the contract could be utilized.

The subcontracts may be prepared by the construction manager or, in a general contracting situation, by the GC subject to approval by the hospital. The architect should also review the contracts.

During the construction process, "common law of the job" could develop, which vitiates the language of even the best contract. That is, if the individuals involved in carrying out the construction project do not follow the terms of the contract, those terms may be waived and therefore not be enforceable at a later date when it is important. All of the procedures established in the contract should be followed or modified in such a fashion that they can be practically followed. "Common law" on the jobsite should be avoided.

Ground Breaking

The hospital may wish to schedule a ground-breaking ceremony in connection with the project. If a bulldozer or other evidence of construction activity are to be on site during the ceremony, this should be confirmed with the construction manager. Often the first two weeks of a construction project are devoted to such site organization and startup activities as moving the construction trailers on-site and finalizing paperwork. Thus, unless explicitly requested, indications of on-site construction activities may not be visible immediately after the official start date of construction.

Should the hospital schedule ground breaking to occur in the midst of actual on-site work, the ceremony logistics must be carefully coordinated with the construction manager so as not to disrupt or interfere with the contractor's work.

Schedule

The construction manager or the general contractor will prepare a detailed schedule of work for each trade category for each week of the entire construction duration. A critical path method (CPM) schedule may be developed from a computer model. The CPM indicates the flow of work from one trade category to another so that proper sequencing is obtained to integrate all components of the work.

Monitoring the construction schedule is extremely important to prevent a shortfall in capitalized interest and this should be emphasized to the construction manager. The on-site construction superintendent should reassess the schedule and report variances to hospital administration on a monthly basis, or more often if needed.

For remodeling projects, it is important to communicate the schedule to the staff of involved departments as well as to the housekeeping and maintenance departments so that preparations may may be made for disruptions and additional cleaning requirements.

Shop Drawings

Shop drawings are detailed plans and installation instructions of engineering systems, components thereof, special building features, and fixed equipment which are supplied by the manufacturers and/or by subcontractors. Shop drawings must be reviewed and approved by the hospital, the architect, and/or engineer to assure that they conform with the specifications of the contract documents. It is important that shop drawings be processed in a timely fashion to avoid construction delays, especially for components of work which are on the critical path.

CHANGE ORDERS

Change orders are revisions to the contract documents which may or may not result in a change in the contract price. Change orders are required to correct errors and omissions on the plans; to respond to changes required by code officials; to remedy problems caused by abnormal field conditions including subsurface conditions and delays due to weather; and to remedy defects in the work.

Change orders may be processed in a number of ways, subject to the terms of the construction contracts. No change orders should be authorized without complete documentation. Certain changes require drawings, sketches, or narrative documentation by the architect. This material may be distributed to the involved contractors for preparation of a fixed price proposal before the owner's approval to proceed. Alternatively, the contractors may be given instructions regarding the change and be authorized to proceed with the work with compensation to be based upon the actual time and materials. In such a case, unit prices and labor rates should be negotiated prior to the notice to proceed.

It is desirable to establish a fixed price before the completion of changes. In certain cases, however, changes affecting work on the critical path must be processed expeditiously in order to maintain the schedule, and a time and materials pricing method is advisable. When time permits, fixed price proposals for change orders should be solicited from the affected contractors, and the construction manager should simultaneously prepare a cost estimate for the change. Then if the contractor's proposed price appears unreasonable, the construction manager should negotiate for a better price. The construction contract should set forth the basis upon which change orders will be performed.

It is essential that change orders be monitored closely during the entire construction period to assure that the contingency is not exhausted prematurely. Change orders resulting from scope changes or discretionary revisions by the hospital staff or the architect should be discouraged in order to preserve the contingency for necessary change orders.

Often, the time period required for contractors to price a change once it has been identified and documented can be lengthy. The hospital should take care to monitor *all* pending changes which are identified but not yet priced in order to correctly assess the adequacy of the remaining contingency. In this regard, the construction manager should provide cost estimates for pending changes at the earliest possible date.

All change orders should be approved by the architect and the owner's representative prior to implementation. Typically the greatest number of change orders occur during the excavation, sitework, foundation, and remodeling stages, since these involve the most unknowns for concealed conditions.

Equipment and Furnishings

The procurement and installation of equipment should be closely coordinated with the construction activities. Fixed equipment is usually covered in the contract documents, but there may be owner-furnished fixed equipment which will be installed by the contractor and which could affect the critical path schedule items. It is important to clarify the cost, schedule, and responsibility for the relocation and hookup of existing fixed or movable equipment to be relocated.

Since much movable medical equipment, especially radiographic equipment, has special requirements regarding size, conduit, mechanical connections, etc., the equipment planning for such items should take place concurrently with design development and construction documents to avoid later change orders. A problem in this regard is that the hospital may wish to defer the equipment selection until the latest date to take advantage of the most current technology available. Another issue which affects the timing of the equipment planning and procurement is the long lead procurement time required for certain items.

Color samples of construction items such as vents, hardware, millwork, etc. should be coordinated with the interior design planning of furnishings and graphics. This could require accelerating the furnishings planning to occur during the design period.

The planning, procurement, and installation of movable equipment and furnishings may be managed by the architect, the equipment planner, or the hospital staff. In preparation for the final delivery of new equipment and furnishings, an inventory and layout for new items and existing items to be relocated should be developed.

Monitoring Project Costs

Although the construction manager is required to monitor and report on construction costs, either the project manager, the financial advisor, or the hospital staff should monitor total project costs during construction.

This is especially important for allowance items for which no firm prices have been established, including change orders, signage, graphics, design fees for change orders, concrete testing, and printing of shop drawings.

During the construction period, all new expenditures and new purchase orders should be allocated to one of the major budget categories (construction, equipment, fees, contingency, administrative expenses) and then to the individual line items within those categories on a monthly basis. It is important to cover all purchase orders as outstanding commitments in the monthly budget evaluation, since unbudgeted items often creep into the project cost and identifying these at the purchase order stage prevents later surprises. The allocation of expenditures among the budget categories is an important task in order to identify specific budget variances.

Another important side of monitoring project costs is the financing category. Capitalized interest must be monitored relative to the schedule and, when capitalized interest is phased, relative to the final actual costs for each building phase.

Interest income on trusteed funds, if applicable, should also be monitored as a use of funds to assure that the expected yield is being attained. Since all funds are invested up front, the initial expected yield can be calculated upon the closing of financing. The final interest earnings could change, however, should the actual project cost draw down lag or be accelerated from the estimated draw down. In such a case either the trustee held investments would be reinvested until needed or early liquidation of investments would be necessary. Then either a surplus or deficit of sources of funds could occur. Should it appear during construction that the actual draw down is differing significantly from the projection upon which the trustee held investments are based, the remaining draw down projection should be revised to guide the trustee in reinvesting funds or preparing for early liquidations.

REPORTS TO GOVERNING BOARD AND ADMINISTRATION

Written reports on the status of the project should be submitted to the governing board and hospital administration on a regular basis. Different levels of detail may be appropriate for the full board, board committee, and hospital administration. The following could be considered for report purposes:

1. Budgeted to projected actual project costs and estimated variance for each major budget category

2. Explanation of budget variances
3. Cumulative expenditures to date and additional commitments for each major budget category
4. Cost and description for major approved change orders
5. Description of major expected change orders not yet approved
6. Description of scope modifications
7. Comparison of estimated to actual expected schedule for each distinct building phase and explanation of major variances
8. Budgeted to expected actual interest income on trusteed funds, if applicable

Warranties

The hospital, with assistance from the architect and the construction manager, should assure that all warranties are properly processed. One problem area that could arise is the start date of the warranty period for types of engineering systems which cannot be tested immediately to assure proper operation. For example, should the building be completed in April, it may not be possible to confirm the proper operation of heating systems until the winter season. In such a case the manufacturer should extend the start date of the warranty until after a full season has elapsed. The warranty period should be specified in the construction contract.

Dedication

As with the ground-breaking ceremony, dedication ceremony activities should be scheduled and the logistics planned to avoid interference with any final construction activities. The dedication should not be scheduled too close to the projected completion date of construction, otherwise a last minute delay in construction cleanup or punchlist activities could cause confusion and interfere with the ceremony.

Generally, dedication activities include a tour through the new facility or addition for major projects. The new facilities will look more impressive if furnishings, signage and graphics are already in place. The schedule for installing these items should be referenced in planning the dedication.

XI

Occupancy and Startup

Upon substantial completion of construction, the architect will conduct a walk-through of the new and remodeled areas and prepare a punchlist of open items to be resolved. The state health department, the fire marshall, and/or local agencies must also conduct a walk-through to inspect the facility for a certificate of occupancy. Depending on the extent of the punchlist and final comments of the agencies, the building may be occupied immediately upon final cleanup. The hospital housekeeping and maintenance departments should have significant time budgeted to ready the facility according to the sanitary requirements of a hospital.

Before patient occupancy, employee orientation sessions should be conducted to familiarize staff with circulation patterns, new operating systems, and new equipment. It may be helpful to seek the facilities planner's assistance in conducting the orientation.

A detailed schedule should be prepared for relocating patients, staff, and supplies in the new or remodeled facilities so that each department is aware of its responsibilities and deadlines. One individual should be designated to plan and coordinate the overall occupancy process. Adequate time should be provided to complete the transition and, depending on the size of the project, it may be necessary to budget downtime and loss of revenue production in certain departments. Whenever downtime is expected, plans must be developed for alternative means of furnishing essential services.

It is inevitable that minor problems will be discovered after the construction project is complete and occupancy occurs. The hospital should maintain strong pressures on the architect and construction manager to resolve all problems in a timely way. Retentions of final payments should be used as leverage where possible to encourage speedy responses to outstanding construction problems.

XII

Conclusion

Efficient and appropriate deployment of capital resources is essential to the operating, financial, and strategic position of a hospital. Capital deployment through investment in construction projects is a process-oriented activity. Key watchwords are timing, coordination, and monitoring. Major guidelines for the process are:

1. The construction project should be part of an overall plan based on the hospital's strategic requirements and long-range financial profile.
2. Preliminary planning is critical. Do not underestimate the required time and resources during the preconstruction stage of the process.
3. Ground rules and procedures should be established for the project development process. Who is in charge? What are the reporting and accountability relationships? What is the communication process? What is the specific scope of work for each project team member? What are the milestone events in the development process?
4. The process is interdisciplinary and interactive. Project team members should not perform their isolated tasks in a vacuum. The design affects the budget, and the financing plan affects the construction schedule. Regular and frequent communication among all parties is essential.
5. Intense hospital involvement is important. Complete delegation to outside parties is not advisable. Conversely, hospital staff members are usually not experts in some or all phases of the

process. An investment in outside technical expertise is often necessary.

6. Scope, budget, and schedule move in tandem. The general parameters, once established, should not be carelessly violated. Stay in bounds and respect the interrelationship of these three key project elements.
7. Expect problems. Search for problems. Foresee problems and evaluate alternative solutions.

Glossary

Administrative expense. A miscellaneous project budget category encompassing soil tests, site survey, builders risk insurance, travel expenses, building permits, etc.

Alternates. Design features which can be separately designed and bid and added to or deleted from the project as the budget dictates.

Authority. A quasi-public state or municipal agency formed as a vehicle for the issuance of tax-exempt debt.

Bid bond. Bond submitted with a construction bid which guarantees the contractor's acceptance of the contract at the bid price.

Block drawings. Architectural drawings indicating blocks of space for departments, primarily developed to identify and test the interdepartmental relationships and functional configurations.

Bond counsel. Legal counsel who represents the municipality or authority that issues tax-exempt bonds or notes and/or opines on the tax-exempt status of the interest on the bonds.

Bond discount. Compensation to the underwriter for managing and selling the bond issue; generally expressed as a percentage of the loan.

Bond rating. Ratings received from the national credit rating services which represent the agencies' evaluation of the ability of the hospital to service the existing and proposed debt.

Building footprint. Overhead view of building outline (size and shape) on site plan.

Capitalized interest. The interest to be paid on the debt during the construction period, nonreimbursable by third-party payers until the facility is in operation, and therefore borrowed as part of the debt proceeds.

Change order. Revisions to the contract documents which may or may not affect the contract price; such revisions may be required to correct plan errors, accommodate unforseen problems, etc.

Certificate of need (CON). Required by federal and state law for capital expenditures in excess of defined amounts, in order to secure, among other sanctions, third-party reimbursement for the capital costs.

Construction fund. Fund held by the bond trustee which contains monies for construction, fees, equipment, contingencies, and miscellaneous project costs.

Construction management (CM). Services provided by a construction firm on behalf of a hospital project, generally including preconstruction consultation and management of the construction process.

Contingencies. Budgeted funds for unanticipated project expenses, including a cost estimating contingency, a design contingency, and a construction contingency.

Critical path method (CPM). A method of project schedule development which indicates flows of work and integrates individual schedule components.

Debt service reserve fund. Reserve fund held by the bond trustee for the term of the loan and used to pay debt service in the event of default. Typically, this equals the maximum annual debt service on the bonds for tax-exempt transactions.

Design development. The design stage at which department managers provide detailed input regarding internal departmental requirements. Locations of casework, outlets, medical gases, etc., are indicated which necessitate interfacing with mechanical, electrical, and plumbing engineers.

Equity contribution. Hospital equity funds designated for use for a given capital project.

Fast track design. A design approach under which plans for early work may be accelerated to begin construction prior to completing the remainder of design.

Feasibility study. Independently forecasted financial statements which indicate the hospital's ability to service its debt.

Front ends. Narrative descriptions accompanying plans and specifications on which bidding contractors calculate their prices, and which include the phasing schedules for construction, payment procedure, required construction methods, etc.

Functional planner. Hospital consultant whose role includes assisting in program development activities, translating projected volume and workloads into space requirements, reviewing existing facilities and code compliances, advising on functional flows within and between departments, etc.

General conditions items. Job site items such as cranes, job site trailer and office equipment, fences, barricades, etc., furnished by the construction manager or the general contractor.

Guaranteed maximum price (GMP). A maximum price provided by one contractor who guarantees to build the building for a not-to-exceed price based on given plans and specifications, which may or may not be complete.

Health systems agency (HSA). A local health care regulatory agency, one of the functions of which is to review and recommend approval of certificates of need.

Interim loan. A source of funds to cover costs during a short time period until the permanent financing transaction can be completed.

Letter of intent. A notice of intent (required in some states) to file a CON application, which is submitted to the health planning and review agencies.

Loan covenants. Provisions in the financing legal documents which impact hospital operations; include rate covenants, insurance requirements, merger/acquisition restrictions, etc.

Millwork. Custom designed built-in cabinetry.

Official statement. The marketing document for a bond issue, which provides potential investors with full information regarding the hospital and the proposed financing.

Owner's representative. This is a term from American Institute of Architects' and Association of General Contractors' documents which refers to the person on the owner's staff designated to approve change orders and pay requests.

Performance bond. Bond submitted by the contractor which guarantees his ability to complete the work for the contract price.

Project manager. Individual responsible for the day-to-day progress of the entire project development and implementation. This may be the same person as the owner's representative. Either the project manager or the owner's representative may be an employee of the hospital or a consulting firm which has contracted with the hospital to provide such services.

Punchlist. Prepared by the architect upon substantial completion of construction to indicate remaining construction items to be resolved.

Purchase contract. Agreement between the hospital and the underwriters stating the price, requirements, and conditions of the bond sale.

Request for proposals (RFP). Sent to architectural, construction, and consulting firms during the project team selection process; delineates project scope, outline of work, proposal format, due date for proposal, etc.

Schematic design. The first architectural layout of spaces within and between departments, including corridors, offices, closets, and other storage areas. The first structural analysis and engineering calculations for the building occur at this stage.

Tax-exempt bonds. Bonds issued publicly or placed privately; issued through a municipality or quasi-public agency at the municipal or state level, the interest of which is exempt from federal or state tax.

Trustee. Commercial bank with which bond proceeds are deposited, the various funds (construction fund, bond payment fund, etc.) are administered, and the debt service payments are received and disbursed.

Underwriter. Investment banking firm whose role includes the structuring of the mechanics of the debt transaction, interacting with the financing authority and rating agencies, advising on the legal financing documents, and marketing the debt.

Supplemental Reading List

PLANNING

Early, W., et al. Avoiding Planning Backlash. *Journal of Business Strategy* 4(2):93-96 (1983).

Gregg, T. E. A Simplified Approach to Capital Project Planning. *Hospital Topics* 57(5):27, 56 (1979).

Gregory, D., et al. The Value of Strategic Marketing to the Hospital. *Healthcare Financial Management* 37(12):16-22 (1983).

Hanvel, E. J., et al. Transitional Planning—An Integrated Approach. *Healthcare Management Review* 8(4):61-67 (1983).

Hayet, L. Remodel or Replace? Deciding What to Do with an Aging Facility. *Trustee* 34(9):27-29, 31-32 (1981).

Kat, B. H. Ranking Hospital Projects—An Allocation Plan. *Hospitals* 54(18):99-100, 102, 104 (1980).

Kutz, G., et al. Strategic Planning in a Restricted and Competitive Environment. *Healthcare Management Review* 8(4):7-12 (1983).

Lee, D. Strategic Planning: Vital for Hospital Long-Range Development. *Hospital and Health Services Administration* 26(4):25-50 (1981).

MacStravic, R. E. Average Life-Cycle Occupancy: A Radical New Approach to Bed Need and Appropriateness Review Decisions. *Health Care Planning and Marketing* 1(1):25-33 (1981).

Nultunson, P. A. Strategy to Implementation—The Planning Process. *Healthcare Financial Management* 37(11):30-34 (1983).

Reeves, P. N. *Strategic Planning for Hospitals.* Chicago: Foundation of the American College of Hospital Administrators, 1983.

Ryan, J. L. Plan, Then Expand. *Health Service Manager* 14(1):12-14 (1981).

Sherlock, T. F. Long-Range Planning: Its Role in Marketing Strategy. *Fund Raising Management* 12(5):54 (1981).

Tichy, N. The Essentials of Strategic Change Management. *Journal of Business Strategy* 3(4):55–67 (1983).

Wyndra, F. T., et al. Outside Consultants: The Pros and Cons of Hiring Them. *Training* 17(6):24–26 (1980).

FINANCIAL PLANNING

Cleverly, W. O. Reimbursement for Capital Costs. *Topics in Health Care Finance* 6(1):25–31 (1979).

Cleverly, W. O., et al. A Survey Report: How Hospitals Measure Liquidity. *Healthcare Financial Management* 37(11):66–72 (1983).

———, et al. Credit Evaluation of Hospitals. *Topics in Health Care Finance* 7(4):1–12 (1981).

Copeland, K., et al. Cost of Capital, Target Rate of Return, and Investment Decision Making. *Health Service Register* 16(3):335–41 (1981).

Hubbard, C. M. Capital Budgeting and Cost Reimbursement in Investor-Owned and Not-for-Profit Hospitals. *Healthcare Management Review* 8(3):7–15 (1983).

Johnson, R. C. Construction Spending Will be Shaped by Competition. *Hospitals* 56(4):89–92, 94 (1982).

Kaufman, K., et al. Financial Manager's Notebook: Strategic Capital Planning, Part 1. *Healthcare Financial Management* 37(3):79–80 (1983).

———. Financial Manager's Notebook: Strategic Capital Planning, Part 2. *Healthcare Financial Management* 37(4):97–98 (1983).

Krystynak, L. F. Prospective Payment for Capital: The Financial Nature of Capital Allowances. *Healthcare Financial Management* 37(10):60–68, 74–76 (1983).

Morrow, J. C. The Debt Capacity Study: A Guide to Determining Affordability. *Texas Hospitals* 37(10):12–14 (1982).

Suver, J. D., et al. Financial Manager's Notebook: Discounted Cash Flow Analysis for Investment Decisions. *Healthcare Financial Management* 37(12):89–90 (1983).

Tierney, T. M., Jr., and L. F. Krystynak. *Capital Analysis and Priority Setting—Capital Formation Concerns in the Health Care Sector.* San Francisco: The Western Consortium for the Health Professions, Inc., 1981.

Vraciu, R. A. Decision Models for Capital Investment and Financing Decision in Hospitals. *Health Services Research* 15(1):35–52 (1980).

Zuckerman, A. Cash Flow Modeling. *Topics in Health Care Financing* 10(1):59–74 (1983).

REGULATORY

American Hospital Association. *Implementation of Certificate of Need for Health Facilities and Services.* Chicago: American Hospital Association, 1975.

Brown, L. D. Common Sense Meets Implementation: Certificate-of-Need Regulation in the States. *Journal of Health Politics, Policy & Law* 8(3):480–94 (1983).

Haag, R. B. Building Codes: Conflicting Requirements Can Be Costly. *Michigan Hospital* 16(6):16–18 (1980).

Herkey, N., et al. Health Planning and Certificate of Need: The Quality Dimension. *Health Policy Quarterly Evaluation and Utilization* 1(4):243–68 (1981).

Hyman, H. *Health Regulation: Certificate of Need and 1122.* Rockville, MD: Aspen Systems Corporation, 1977.

Joskow, P. L. Certificate of Need Regulation. In *Controlling Hospital Costs*, edited by P. L. Joskow, 77–99. Boston: Massachusetts Institute of Technology Press, 1981.

Kuntz, E. F. Planning Laws Survive Cuts, Face Changes. *Modern Health Care* 12(3):86 (1982).

Reagan's Budget Would Foreclose on PSRO's, HSA's. *Medical World News* 23(5):11, 15 (1982).

Salkever, D. S., and Thomas W. Bice. The Impact of Certificate of Need Controls on Hospital Investment. *Milbank Memorial Fund Quarterly* 54:185–214 (1976).

U. S. Department of Health and Human Services. *Minimum Requirements of Construction and Equipment for Hospital and Medical Facilities.* Washington, DC: U. S. Department of Health and Human Services, 1978.

Public Health Service, Bureau of Health Planning. *Certificate of Need Programs: A Review, Analysis and Annotated Bibliography of the Research Literature.* Washington, DC: Government Printing Office, 1978.

Design Development

Bentivegna, P. L. Redesign Saves Project. St. Joseph's Hospital Meets HSA-Approved Budget/Fulfills Space Requirement. *Hospitals* 57(4):85–88 (1983).

Bobrow, M., and P. Van Gelder. The Best in Hospital Architecture. *Hospital Forum* 23(4):13–17 (1980).

Ford, W. R. Hospital Space Planning—The Role of the Financial Manager. *Healthcare Financial Management* 37(11):14–18 (1983).

Frommelt, J. J., et al. Construction Budget: The Financial Manager as Part of the Planning Team. *Healthcare Financial Management* 37(11):20–26 (1983).

Hardy, O. B. and L. Lammers. *Hospitals—The Planning and Design Process.* Rockville, MD: Aspen Systems Corporation, 1977.

Haviland, D. *Managing Architectural Projects: The Effective Project Manager.* Washington, DC: American Institute of Architects, 1981.

Hendrickson, W. W. The Space Challenge. Hospital Suggests Space Redesign, Uninterrupted Patient Care. *Hospitals* 57(4):102–4 (1983).

Kuntz, E. New Designs Cut Construction Costs. *Modern Health Care* 10(2):60–62 (1980).

McLaughlin, H. Design-CM Team Needs Hard Coaching from Hospitals. *Modern Health Care* 11(2):94, 96 (1981).

Omens, R. S. Building and Remodeling: What You Need to Know and Do before Construction Begins. *Hospital Forum* 23(4):10–12 (1980).

Porter, D. *Hospital Architecture: Guidelines for Design and Renovation.* Ann Arbor, MI: AUPHA Press, 1981.

Sprague, J. G. The Art of Estimating. Accurate Estimates Depend on Project Quality and Quantity. *Hospitals* 57(4):80-82 (1983).

Taylor, W. J. Expansion within Bounds: Wrestling with Space and Budget Limitations. *Hospitals* 57(4):95-96 (1983).

Ziskind, D. M. Evolutionary Design. Taking a Project through Concept, Program, and Detail. *Hospitals* 57(4):107-8 (1983).

PROJECT FINANCING

Blaes, S. M. Tax-Exempt Bond Financing Considerations for Catholic Hospitals. *Hospital Progress* 63(2):36-41, 52 (1982).

Clairborn, S. A., et al. Floating Interest Rate Bonds—For Some Hospitals, an Attractive Alternative. *Hospital Financial Management* 35(10):59-61 (1981).

Collins, J. Dealers, Hospitals May Benefit from Creative Bond Financing. *MPS: Medical Products Salesman* 12(7):26-27, 29 (1981).

Friedlander, G. D. Complex Factors Determine Yield in Hospital Bond Issues. *Hospitals* 55(15):81-86 (1981).

Hee, D. L., et al. Tax-Exempt Bonds: Operational Statistics and Bond Ratings. *Healthcare Financial Management* 37(10):52, 54, 56 (1983).

Hospital Revenue Bonds: How They're Rated. *Hospital Financial Management* 11(9):47 (1981).

Kalick, L. L. et al. ERTA, TEFRA and the Tax Exempt Entity: Financing Projects with Outside Capital. *Healthcare Financial Management* 37(4):54 (1983).

Lightle, M. A. Pooled Equipment Financing: An Attractive Alternative. *Health Law Vigil* 4(13):3-4 (1981).

Lucksinger, T. S., et al. Capital Management: Avoiding the Pitfalls in Long Term Financing. *Hospital Financial Management* 35(10):46-48, 50, 52 (1981).

Mullner, R., et al. Debt Financing: An Alternative for Hospital Construction Funding. *Healthcare Financial Management* 37(4):18-20, 24 (1983).

New Revenue Sources and Tax Exempt Status. *Hospital Financial Management* 11 (10):59-61 (1981).

Refinancing Capital Debt. Tax-Exempt Bonds: Tool for Containing Costs. *Cost Containment* 1(18):3-6 (1979).

Rosenfeld, R. H. Tax-Exempt Hospitals Explore New Ways to Attract Equity. *Hospital Financial Management* 35(10):59-61 (1981).

Rum, J. The Outlook for Tax-Exempt Financing. *Trustee* 36(12):29-30 (1983).

Taddey, A. J. and G. Gayer. Results of Study: Uses and Effects of Hospital Tax-Exempt Financing. *Healthcare Financial Management* 36(7):10-13, 16, 20-25 (1982).

Vraciu, R. A. Three Rules for Selecting Capital Financing Options. *Hospital Financial Management* 34(4):38-46 (1980).

What's a Hospital Feasibility Study and Who Needs It? *Hospital Financial Management* 11(9):47 (1981).

Zaretsky, H. W. Capital Financing in the 1980s. *Issues in Health Care* 2(1):58-59 (1981).

Fundraising

All-Out Fundraising. *Profiles in Hospital Marketing.* (2):78–81 (1981).
Harman, W. J. Strategic Plans Increase Capital Campaign Success. *Fund Raising Management* 12(6):38–41 (1981).
Panas, J. *Megagifts. Who Gives Them, Who Gets Them.* Chicago: Pluribus Press, Inc., 1984.

Construction

Campbell, S. R. Procurement of Major Equipment. *Hospital Materials Management Quarterly* 2(3):17–30 (1981).
Hornell, J. Do's and Dont's for a Successful Construction Project. *Trustee* 34(9):20–21 (1981).
Jones, R. J. Alternative Construction Contracts Save Hospitals Thousands of Dollars. *Modern Healthcare* 10(6):74–75 (1980).
Miller-Jones, M. The Construction Manager Can Be the Hospital Trustee's Ally. *Trustee* 36(2):33–34, 36 (1983).

Occupancy

Jaye, D. R., Jr. St. Joseph's Hospital, Marshfield, WI: Adding Space, Moving 170 Patients Takes Enthusiasm, Lists, Rehearsals. *Hospital Progress* 60(12):28–30 (1979).
Morgan, K. J. Coping after Relocation in a New Hospital. *Times* 22(7):13–14 (1981).

Index

About the Authors

DEBORAH J. ROHDE is Senior Vice President of Health Facilities Corporation, a health care consulting firm based in Northfield, Illinois. Ms. Rohde has provided project management and financial advisory services to 30 hospital construction projects totaling $700 million. Her hospital advisory experience also includes negotiation of architectural and construction contracts, negotiation of loan documents, business planning, joint venture studies, and medical office building development. Ms. Rohde has lectured at hospital and professional association seminars on strategic capital planning, the capital financing process, and managing capital construction projects. Ms. Rohde has a B.A. degree from Eureka College and has pursued coursework in finance at the University of Chicago Graduate School of Business.

LAWRENCE D. PRYBIL is Executive Vice President, Daughters of Charity Health System, East Central, Inc., based in Evansville, Indiana. At the time this book was written, he was Vice President for Administration, Sisters of Mercy Health Corporation, where he was on the Task Force on Construction Planning and Management. He has served as an Instructor and Research Associate at The University of Iowa; as Professor and Chairman, Department of Health Administration, at the Medical College of Virginia—Virginia Commonwealth University; and as Adjunct Professor at The University of Michigan's School of Public Health. He has served on the governing boards of hospitals, state hospital associations, and multi-unit hospital systems. His publications include several journal articles, book chapters, and research reports, and he has served on numerous editorial boards. He has a B.A. in Liberal Arts and a M.A. and Ph.D. in Hospital Administration from The University of Iowa.

WILLIAM O. HOCHKAMMER is a partner of Honigman Miller Schwartz and Cohn in Detroit, Michigan. Since obtaining his J.D. from the Northwestern University School of Law in 1969, he has taught at the Detroit College of Law and the Wayne State University School of Law. His other professional activities include being past president of the Michigan Society of Hospital Attorneys, past chairman of the Committee on Legal Services, Catholic Health Association, and counsel to numerous health care organizations. Hochkammer received his B.A. from Lawrence University.